Welcome to the Great Unraveling: Navigating the Polycrisis of Environmental and Social Breakdown
Copyright © 2023 by Post Carbon Institute.
All rights reserved. Published June 2023.

Authors: Richard Heinberg and Asher Miller.
Editorial Committee: Rob Dietz and Daniel Lerch.

Cover image photo: "The Great Unraveling" by Michele Guieu; commissioned by Post Carbon Institute.

Visit postcarbon.org/wgu-report for further resources and errata notes.

Post Carbon Institute's mission is to lead the transition to a more resilient, equitable, and sustainable world by providing individuals and communities with the resources needed to understand and respond to the interrelated ecological, economic, energy, and equity crises of the 21st century.
postcarbon.org | resilience.org

Post Carbon Institute
P.O. Box 1143
Corvallis, Oregon 97339 - USA

Contents

Preface iii

I. **Understanding the Great Unraveling** 1
    A. Overview 1
    B. The Polycrisis Defined 2
    C. The Origins of Our Predicament 5

II. **Environmental Unraveling** 9
    A. Global Warming 11
    B. Biodiversity and Habitat Loss 12
    C. Soil Loss and Degradation 12
    D. Water Scarcity 13
    E. Chemical Pollution 13
    F. Resource Depletion 14
    G. Summary: Feedbacks and Underlying Drivers 15

III. **Social Unraveling** 17
    A. Poverty 18
    B. Inequality 18
    C. Racism and Other Forms of Discrimination 19
    D. Antisocial Responses to Scarcity 19
    E. Authoritarianism 21
    F. Impacts of Technological Change 21
    G. Summary: Feedbacks and Underlying Drivers 22

IV. **Pulling on the Threads: The Interaction of Environmental and Social Drivers** . 25
    A. The New Reality 25
    Sidebar: The Spectrum of Destabilization > Breakdown > Collapse 27
    B. Destabilization, Feedbacks, and Conclusions 28

V.	**Weaving a New Tapestry: How to Respond to the Great Unraveling**	31
	A. Navigating the Liminal Space	32
	B. Recognizing Obstacles to Systemic Change	32
	C. Getting It Right or Getting It Wrong	39
VI.	**What You Can Do**	45
	A. Informational Competence	46
	B. Emotional/Psychological Resilience	47
	C. Practical Personal Steps for Building Resilience	49
VII.	**Summary and Takeaways**	51
	Endnotes	54
	Image Credits	63

Preface

During the 20th century, and especially the latter half of the century, humanity's increasing adoption of fossil fuels as sources of cheap and abundant energy enabled rapid industrialization. The result was a massive increase in nearly all human activities and their ecological and social impacts, a process that has been called the *Great Acceleration*.[1] The first two decades of the 21st century saw a new phase of the Great Acceleration, with wars fought over the last sources of cheap oil, expensive and destructive exploitation of remaining natural resources, the massive use of debt and speculation to expand energy production and maintain economic growth, and the arrival of environmental and social impacts too overwhelming for even the world's wealthiest and most powerful people and nations to ignore.

Now, in the 2020s, the Great Acceleration is losing steam and shows signs of reversing direction. Thought leaders and policy think tanks have invented a new word—*polycrisis*—to refer to the tangles of global environmental and social dilemmas that are accumulating, mutually interacting, and worsening. The central claim of this report is that the polycrisis is evidence that humanity is entering what some have called the *Great Unraveling*[2]—a time of consequences in which individual impacts are compounding to threaten the very environmental and social systems that support modern human civilization. The Great Unraveling challenges us to grapple with the prospect of a far more difficult future, one of mutually exacerbating crises—some acute, others chronic—interacting across environmental and social systems in complex ways, at different rates, in many places, and with different results.

Welcome to the Great Unraveling is intended to help the general public—but particularly academics and researchers, environmental and social justice nongovernmental organizations and their funders, and the media—recognize what the Great Unraveling is, what it means for both human civilization and the global ecosystem, and what we can do in response. The paper calls attention to four main things:

1. the alarming, rapidly changing environmental and social conditions of the Great Unraveling;

2. the need to grapple with complexity, uncertainty, and conflicting priorities—hallmarks of the Great Unraveling;

3. the need to maintain social cohesion within societies and peaceful relations between them during the Great Unraveling, while implementing key changes in collective behavior and managing the negative consequences of past failures to act; and

4. the personal competencies that are needed to understand what's happening during the Great Unraveling and to respond constructively, primarily by building household and community resilience for this precarious time.

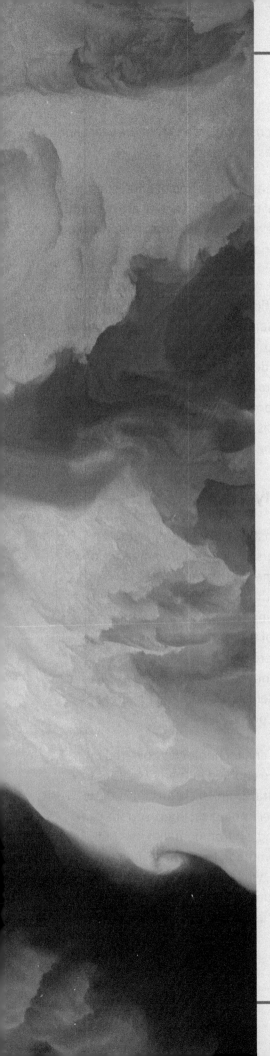

I. Understanding the Great Unraveling

A. Overview

For decades, environmental scientists have warned that ongoing global industrial expansion and population growth based on ever-increasing fossil fuel consumption would eventually trigger a series of linked, snowballing crises.[3] During the same period, social scientists have identified worsening racial, ethnic, and economic inequality and exploitation as drivers of societal destabilization. These environmental and social trends, it was warned, together could result in widespread failure of global ecosystems and industrial societies.

To be sure, efforts have been made in recent decades to address worsening environmental and social crises. On the environmental front, policy makers—pushed by a movement of organizations and activists—have cleaned up rivers, helped bring some species back from the brink of extinction, slowed deforestation, regulated harmful substances, and reduced local greenhouse gas emissions. In the 1980s, a coordinated, successful program to halt the destruction of Earth's atmospheric ozone layer showed the ability of the international community to tackle global environmental challenges.[4] On the social front, the past century has seen the expansion of worker and voter rights, improvements in material conditions for billions of people, and concerted efforts to eliminate discrimination against women and minority groups. However, in spite of these steps, leaders have for the most part failed to address the underlying causes of environmental and social unraveling.

Since the start of the COVID pandemic, Russia's invasion of Ukraine, and the resulting disruption of multiple global supply chains, the term *polycrisis* has entered our global vocabulary. The World Economic Forum's *Global Risks Report 2023*[5] defines the term as "a cluster of interdependent global risks [that] create a compounding effect, such that their overall impact exceeds the sum of their individual parts." Scholars from a range of disciplines (including Columbia University historian Adam Tooze[6]) have written about the emerging polycrisis, and think tanks such as Cascade Institute and Omega Institute have published papers and reports on it.[7]

Welcome to the Great Unraveling seeks, first, to trace the ultimate sources of the polycrisis, explaining why it is the inevitable product of global institutions and trends of the past decades (including persistent economic growth, rapid population increase, capitalism, and our systemic reliance on fossil fuels); and second, to explore the meaning of this crucial juncture in human affairs and what it requires of us. With the emergence of the polycrisis, humanity has entered an age of consequences. The authors of this report have adopted the term *The Great Unraveling* (coined by Joanna Macy[8]) to describe what comes next, and in order to emphasize that:

- the polycrisis was and is inevitable, given its historical roots;
- as a result of continuing failure to address those roots, multiple environmental and social crises will intensify in the years and decades ahead; and
- as we enter this period of unraveling, new ways of thinking and acting are required at all levels of society, from policy makers to ordinary citizens.

Fifty years ago, we (i.e., humanity) faced a difficult choice: dramatically shift our trajectory or eventually suffer the results. Now we must both contend with the unavoidable environmental and social costs of delay in addressing the root causes of unraveling, *and* finally get down to the hard work of transforming the underlying systems that are driving crisis upon crisis, if we are to have any hope of averting the worst.

B. The Polycrisis Defined

The Cascade Institute notes that "a global polycrisis occurs when crises in multiple global systems become causally entangled in ways that significantly degrade humanity's prospects. These interacting crises produce harms greater than the sum of those the crises would produce in isolation, were their host systems not so deeply interconnected."[9] There is, as we will see, abundant evidence that humanity is already in the throes of a polycrisis.

In the modern era, technologies powered by fossil fuels have enabled humanity temporarily to exceed natural limits to expansion and exploitation. These technological interventions came with human and environmental costs—but benefits were immediate for those in privileged circumstances, while costs were externalized to those less fortunate, and to future generations and other species. At the dawn of the 20th century, farmers in many nations were facing declining soil fertility, but artificial fertilizers made with fossil fuels enabled burgeoning harvests—while also polluting air, soil, and water, and creating "dead zones"[10] in oceans. Fishers and foresters used fossil energy to harvest renewable resources faster and more thoroughly than was previously possible, supplying society with more food and raw materials while damaging ecosystem vitality and making future harvests problematic.[11] Miners, confronting the exhaustion of high-grade ores, used more fuel to extract and process lower-grade ores[12]—creating more toxic waste and leaving less ore in the ground for future generations. The cheap and rapid transport of goods erased local scarcities, making it possible for people to live nearly anywhere—expanding zones of human livability and in the process increasingly robbing other species of habitat.[13]

Rapid industrialization fed on cheap, abundant fossil energy, which fueled an acceleration of nearly all human activities and their impacts. Because we have been born into it, this decades-long Great Acceleration is assumed by many people to be a normal state of affairs—but it is arguably the most anomalous time in human history.[14]

Now, as the brief fossil fuel era draws to a close and an era of consequences begins, the trends of overall expansion in human population and economic activity are slowing,[15] while the negative impacts of growth to date are compounding.

Most discussion about human impact on the environment centers on climate change, which could undermine food systems and make large swaths of the planet uninhabitable. However, other kinds of human-caused environmental unraveling likewise have the potential to bring civilization to its knees; these include topsoil depletion and degradation, loss of habitat and biodiversity, and the saturation of the global environment with chemicals that disrupt reproduction in humans and other animals.[16] (We discuss environmental impacts in more detail in section II below.)

Meanwhile, the consequences of economic acceleration (including, ironically, its inevitable slowing) are leading to social unraveling. During rapid industrialization, wealth inequality and economic exploitation (both between and within nations) was often tolerated or excused with the assumption that economic growth was a tide that would eventually lift all boats. In reality, economic benefits were very unevenly distributed. As acceleration stalls, many of those who were left behind and who supplied the industrial system with cheap labor and raw materials are responding with political radicalization, while those who benefitted disproportionately are fighting to keep their advantages. (See section III below.)

Societies have confronted challenges before, but these tended to be localized. Today, civilization is global. Humanity has become a "Superorganism" due to instantaneous globe-spanning communication and rapid long-distance transportation enabling large-scale movement of raw materials and finished goods.[17] A typical smartphone may be designed on one continent, draw raw materials from three others, get assembled on a fifth, and be shipped to a sixth for ultimate sale.[18] The impacts of human activities are also globalized as never before, with greenhouse gas emissions from one country affecting the climate for people on the other side of the planet—often, people who themselves generate minimal emissions.

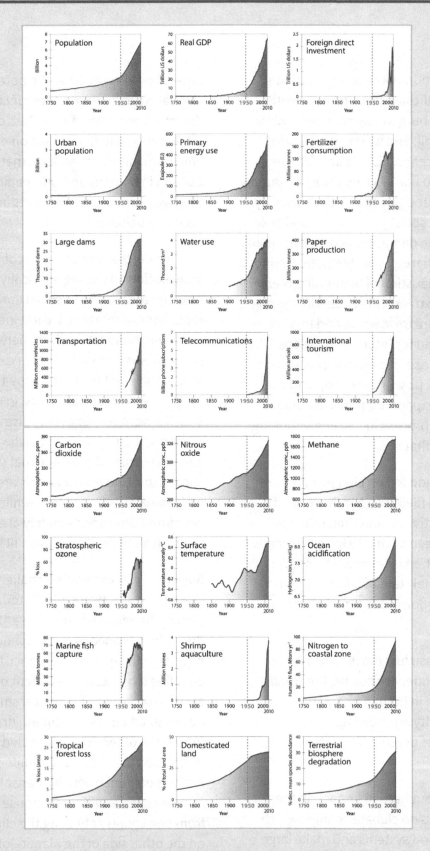

Figure 1. The Great Acceleration: Socio-economic trends and Earth system trends, 1750-2010

Source: Steffen et al., "The Trajectory of the Anthropocene: The Great Acceleration," (2015).

Global industrialization, driven largely by neoliberal economic policies and the growing influence of multinational corporations, has led to a focus on economic efficiency and specialization. Among the lessons of the COVID-19 pandemic is that our global economy is vulnerable to shocks or disruptions anywhere in the system. Goods and services that meet critical needs were once produced closer to home and at multiple locations, but with global integration we've lost the resilience that redundant local production and distributed inventories formerly provided. Scarcity, delays, and inflated costs of goods of all kinds—triggering political upheaval and humanitarian crises—are the inevitable result.

The pandemic brought other useful lessons: that we humans can sometimes adapt quickly when we need to, that social cohesion is required for dealing with existential challenges, and that clear and factual communication from uncorrupted leaders is essential. However, we also learned that most leaders fail to anticipate systemic threats, and that some are driven more by politics, power, and ideology than by reason or by concern for the common good. Further, we learned that the public's patience for sacrifice and behavioral change is limited, and that most people yearn for circumstances to quickly return to normal, even if that "normal" condition is inherently unsustainable. Finally, we discovered that systemic threats, such as pandemics or climate change, can only be countered using *systems thinking*, which enables us to explore the links between seemingly discrete sectors of nature and human society.

The central metaphor we have chosen for this report is *unraveling*. It calls to mind the image of a tapestry—a fabric woven with intention over time—that is beginning to fray. Likewise, global civilization is woven from multiple threads, in the forms of intertwined natural and human systems, which are now individually threatened, continually interacting, and losing mutual coherence.

When a tapestry starts to unravel, at first only a few threads may be lost. Over time, the integrity of the whole declines. Our civilization appears to be in the early stages of unraveling, and maintaining it will require increasing intention and effort. If humanity is unable to muster a shared and sustained intention to do what is necessary to repair and reweave the fabric that binds us together, then that fabric will cease to hold strong. Infrastructure will decay. Alliances will fail. Nations will be rent by tribalism. Environments will cease to be habitable. Altogether, this unraveling will present a far greater challenge than the recent global pandemic. It will affect everyone, it will persist and worsen, and, if we do nothing, it may not be humanly survivable.

C. The Origins of Our Predicament

Understanding the origins of the Great Unraveling requires a historical perspective rooted in the latest, most reliable findings of social science. Of particular relevance is the study of societal dynamics—how and why societies sometimes grow and other times collapse.

Today's social evolution theorists view human societies as systems with *enablers* and *limits*.[19] The chief enabler of, and limit to, societal growth is available energy.[20] History reveals how people have found ingenious ways of gathering energy (whether from food, firewood, wind, flowing water, or fossil fuels) and harnessing it to achieve goals. Technology leverages energy to accomplish tasks ranging from building homes to sending messages to defending against hostile neighbors.

War was another key factor in societal growth and evolution, providing a constant incentive for expansion so as to outcompete other societies for resources and energy, while also generating more cooperation and technological innovation within societies.[21]

During the past 11,000 years (a period of stable, favorable climate known as the Holocene), the human enterprise grew in fits and starts punctuated by periods of stagnation or retrenchment. Roughly 6,000 years ago, grain agriculture—which provided food energy in forms that could be stored and taxed—led to the rise of early states with full-time division of labor and stark hierarchies, extending from kings to soldiers to peasants to slaves.[22] While many Indigenous peoples across the globe continued to live in decentralized and non-hierarchical communities, increasingly complex agrarian societies came to dominate ever-larger regions of the planet. Three millennia ago, war pushed the scale of organization of societies to another level with the emergence of empires, which brutally ruled enormous geographic regions, but guaranteed non-enslaved citizens certain freedoms and rights.[23] Five hundred years ago, trade competition among European nations, plus advances in naval technologies and weapons, led to a rapid global colonial takeover of much of the world by empires centered in England, Spain, France, Portugal, and the Netherlands.

Still, all societies, from the richest to the poorest, shared an energy regime based on recent flows of solar energy, which fed the plants that humans and other animals consumed, and propelled the wind currents that transported people and materials across great distances.

An enormous expansion of Europe's wealth, derived largely from exploiting the people and resources of its colonies, drove developments in the 17th and 18th centuries that laid the groundwork for a new energy regime: government support for private ownership of natural resources and capital investment, and a self-reinforcing feedback loop between scientific research and technological innovation. Then, in the 19th and 20th centuries, capitalism and rapidly evolving technology made it possible to extract and use fossil fuels on a society-changing scale.

The amount of energy that could be unleashed from these fuels was vast, representing tens of millions of years' worth of ancient sunlight.[24] From a quantitative standpoint, it was far and away the biggest energy breakthrough in human history. Soon coal, then oil and natural gas supercharged the processes of mining, manufacture, transport, agriculture, and scientific discovery. Society was reshaped by the ending of institutionalized slavery (since coal generated more wealth for industrialists than the labor of enslaved persons did for plantation owners), as well as by rapid urbanization and the burgeoning of the middle class. Industrial food production, together with sanitation chemicals and pharmaceuticals made with fossil fuels, enabled human population to grow at the fastest (by far) sustained rate in history—rising from one billion in 1820 to eight billion in roughly two centuries.[25]

By the mid-20th century, growth had become normalized in industrial nations. Once economists understood that the Great Depression had resulted partly from overproduction of goods, leaders of government and industry collaborated

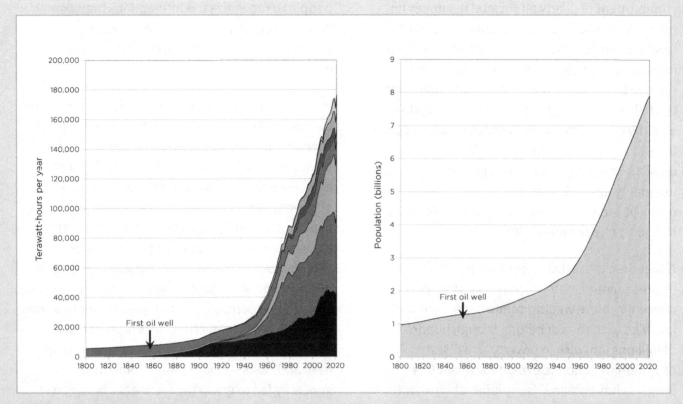

Figure 2. Global total energy consumption growth (left) and human population growth (right), 1800-2021

Charts by J. David Hughes. Data: Hannah Ritchie et al., "Energy" (2022) and Our World in Data, "Population by world region, including UN projections" (2023).

to develop a new socioeconomic system—*consumerism*.[26] Advertising encouraged more consumption while consumer credit enabled it, thereby generating more profits, jobs, returns on investments, and tax revenues. The economy was now a "thing" to be measured via gross domestic product (GDP) and controlled via government-influenced interest rates, with growth as the continual, overarching goal.

However, growth did not benefit everyone equally. From the times of early state societies, inequality had tended to increase until checked by revolt, war, pestilence, or economic collapse.[27] In the fossil fuel era, as overall wealth grew faster than at any previous time, there was the potential for far greater inequality and thus more social instability. Labor strikes roiled industrial economies in the late 19th and early 20th centuries, leading eventually to communist revolutions in some countries, and in others to "leveling" reforms including prohibition of child labor, establishment of the 40-hour work week, higher taxes on the wealthy, and redistributive programs to prevent the poor from falling into utter destitution. Two World Wars, which required the commitment of nearly all societal resources for the nations involved, also resulted in economic leveling, temporarily but substantially reducing inequality. However, during ensuing times of relative peace and stability, and particularly after the fall of the Soviet Union in 1991, the world's wealthy tended to use their social power to beat back leveling efforts. During the last 30 years, driven by neoliberal economic policies, economic inequality has increased to the point where fewer than 100 individuals control as much wealth as the poorer half of humanity.[28]

Altogether, the last couple of centuries have seen unprecedented material progress. More people now enjoy more wealth, comfort, security, and knowledge than ever before, though hundreds of millions languish in poverty as a result of systemic economic exploitation. But this progress has entailed rising vulnerability—with both a social and an environmental aspect.[29] Socially, the fossil-fueled industrial enterprise is rife with human exploitation, with enormous temporary benefits for a fortunate few and permanent deprivation for multitudes of others. Extreme inequality begets political instability. Environmentally, the combustion of fossil fuels is altering the global climate in ways that threaten the biosphere; further, a far larger population consuming resources at a faster per-capita rate translates to more resource depletion, pollution, and damage to nature.[30]

Unfortunately, this vulnerability tends to be hidden from the view of policy makers and ordinary citizens—especially those in relatively rich countries that have benefitted disproportionately from industrialization. Urbanization and labor specialization create wealth and social power, but they also often make it harder for people to recognize the interactions between human and natural systems, normalizing a situation that is in fact unprecedented and perilous.

Hence most policy makers are effectively asleep at the wheel. They are aware of the threat of climate change, but typically see it as a technical problem having to do with carbon emissions, rather than as a systemic predicament intertwined with economic expansion, environmental and human exploitation, and population growth. Leaders (progressive ones, that is) hope for a future of "green growth" based on renewable energy and in which social problems can be solved by further industrial expansion. But few leaders understand what would be required for a comprehensive energy transition, or the further environmental risks and costs it would entail.[31] And they generally fail to grasp the feedbacks between societal vulnerabilities related to the environment and those stemming from social inequality.[32]

Now, the time available to *prevent* unraveling has elapsed. We have only a brief period in which to find ways to hold together the most vital of threads, while contending with the cascading consequences of misguided human actions to date. Doing this successfully will require the development of cooperative, adaptive skills and behaviors across the social spectrum.

II. Environmental Unraveling

In recent years, environmental scientists have sought ways to quantify the kinds and degrees of negative human impact on the Earth's biosphere, and to make this information comprehensible to policy makers and the public at large. Two useful ways of framing the relevant data are planetary boundaries and ecological footprint analysis.[33]

The nine currently identified planetary boundaries define a "safe operating space for humanity." According to Earth system scientists, "transgressing one or more planetary boundaries may be deleterious or even catastrophic due to the risk of crossing thresholds that will trigger non-linear, abrupt environmental change within continental-scale to planetary-scale systems."[34] Currently five boundaries (climate change, biodiversity loss, land-system change, biogeochemical flows, and "novel entities") have been breached and present imminent catastrophic risk.[35]

Ecological footprint analysis measures human demand on the biosphere, i.e., the burden we impose on nature to support people or an economy, as a percentage of Earth's regenerative capacity. Analysis shows the Earth has been in "overshoot"—where humanity is using resources at a pace that ecosystems cannot renew, and generating waste at a pace that ecosystems cannot absorb—since the 1970s.[36] Humanity now lives unsustainably by depleting natural resources, using at least 1.7 times more than what the planet's ecosystems renew. This overshoot is not equally distributed. Some countries, and some households

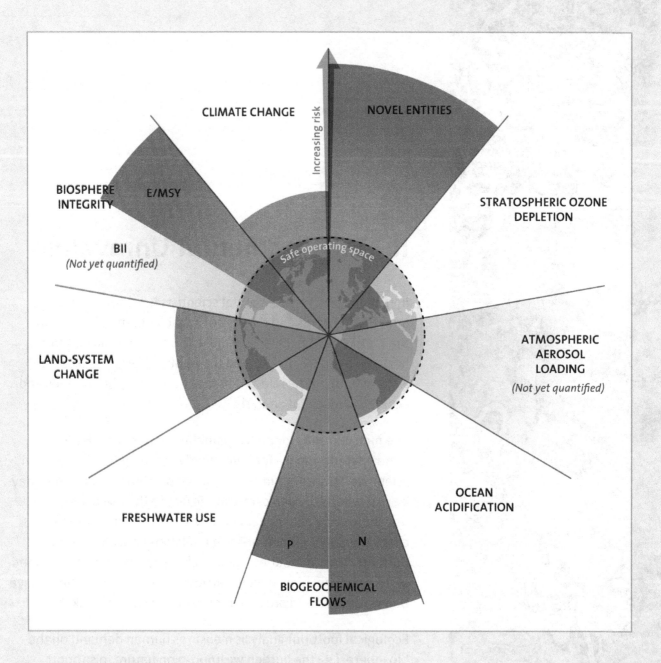

Figure 3. The planetary boundaries framework

E/MSY = Extinctions per million species per year; BII = Biodiversity Intactness Index; P = Phosphorus; N = Nitrogen. Image credit: Stockholm Resilience Centre, based on analysis in Steffen et al, "Planetary boundaries" (2015) and Linn Persson et al., "Outside the Safe Operating Space of the Planetary Boundary for Novel Entities" (2022).

within countries, consume a far greater share of resources than others. If everyone in the world consumed what an average person in the United States consumes, we would need five Earths to support the population over the long term. In contrast, if we consumed at the same rate as the average Tanzanian, we would only need 70 percent of one Earth.[37]

In recent years, scientists have issued increasingly dire warnings with regard to accelerating environmental impacts. Most recently, in a report in *Frontiers in Conservation Science*, 17 authors referenced more than 150 studies to underscore their conclusion that "The scale of the threats to the biosphere and all its lifeforms—including humanity—is in fact so great that it is difficult to grasp for even well-informed experts."[38]

The repeated warnings by scientists center on the following six environmental crises.

A. Global Warming

Human-caused climate change is largely the result of humanity burning fossil fuels, cutting down forests, and farming livestock.[39] It is worsened by continuing population and consumption growth, which entail more energy usage, more landscape alterations, and more food production.

The 2022 Intergovernmental Panel on Climate Change (IPCC) Sixth Assessment report[40] concluded that the global climate has already warmed by 1.2° Celsius (2.2° Fahrenheit) above pre-industrial levels and is set to reach 1.5°C (2.7°F) warming as soon as 2027.[41] Temperatures and sea levels will continue to rise. And the world's "carbon budget" (the amount of carbon that can be emitted in the future without triggering catastrophic warming) will be exhausted within a decade at current emissions rates.

In 2019, a statement by 11,000 scientists warned that the world will face "untold suffering due to the climate crisis" unless major changes are made.[42] The threats are startling:

- Increases in temperatures,[43] changes in precipitation patterns, an increase in number and severity of extreme weather events, and a decline in water availability may all result in reduced agricultural productivity.[44] Access to food (due to rising prices), and food quality, may be severely and widely impacted.

- Rising seas could add 15 inches to sea levels by 2100, displacing up to 630 million people.[45]

- In coming decades, some regions of the world may become so hot as to become uninhabitable by humans, with 3 billion seasonally experiencing "near-unlivable" temperatures.[46]

- Worsening droughts and floods will likely drive increasing numbers from their communities, leading to waves of refugees and migrants (up to a billion by 2050[47]) overwhelming nations and regions that are less impacted.

So far, the scale of climate change mitigation efforts has been a small fraction of what would be required to keep warming to the UN-agreed goal of 1.5°C.[48] Meanwhile, geoscientists warn that cascading, self-reinforcing climate feedbacks may already have been triggered that would impair humanity's chances of significantly slowing global warming.[49]

B. Biodiversity and Habitat Loss

Wild nature is suffering an accelerating die-off both of numbers of species, and numbers of individuals within most species. This die-off is being caused by human land use that destroys habitat; pollution from pesticides, herbicides, fertilizers, and other industrial chemicals; invasive species often transported by humans; and human-caused climate change.

An estimated one million species are at risk of extinction, many within decades.[50] Rates of extinction, currently estimated to be 10,000 times the "normal" rates, are not yet comparable to those seen in mass extinction events earlier in Earth history when up to 97 percent of species disappeared. However, scientists (depending on the methodologies selected for their analyses) suggest that the likelihood of a mass extinction event ranges from possible (at best) to inevitable (at worst).[51]

Meanwhile, the number of individuals within most species is also plummeting. The 2020 WWF Living Planet report determined that the average population size of vertebrates (including mammals, fish, birds, amphibians, and reptiles) had declined by 69 percent in the past five decades.[52] Up to 70 percent of flying insects have also disappeared.

The potential impact to humans is difficult to quantify, as many *ecosystem services* (for example, the pollination of food crops by bees) are simply taken for granted. A World Economic Forum report in 2020 named biodiversity loss as one of the top threats to the global economy.[53]

C. Soil Loss and Degradation

Soil is the basis of agriculture, and hence of civilization itself. Unfortunately, settled human societies have a long history of treating their soils neglectfully, leading to erosion, salinization, desertification, and the depletion of essential plant nutrients such as nitrogen and phosphorus. It is possible that such patterns of neglect even contributed to the collapse of early civilizations.[54] Today, these same disturbing trends are being intensified and broadened by population growth, industrialized food production, globalized markets, and the use of artificial fertilizers made with fossil fuels.

Soil forms slowly, averaging growth of one inch per century, but soil loss due to human activity is happening much faster. Globally, the annual loss of 75 billion tons of soil costs the world about $400 billion per year in lost agricultural productivity.[55] Extreme rainfall and flooding, which are becoming more frequent and intense due to climate change, increase soil erosion and nutrient loss.[56] The greatest impacts from such events are predicted to occur in Sub-Saharan Africa, South America, and Southeast Asia.[57] If erosion and depletion continue to impact soils to the point that fertile land becomes scarce, it's hard to fathom how dollar costs could account for what was lost.

Currently, the depletion of essential plant nutrients from soil is addressed by applying commercially produced nitrogen, phosphorus, and potassium fertilizers. In addition to pollution issues associated with these fertilizers, there are also potential supply constraints with two of the three. Nitrogen (ammonia) fertilizers are currently made using natural gas, which is depleting and polluting.[58] Unlike nitrogen fertilizers, phosphorus soil amendments are not manufactured; they are mined. Although phosphorus is relatively plentiful in Earth's crust, only phosphate in rock deposits can be extracted economically, and such deposits are rare and quickly depleting, with price volatility looming on the horizon.[59]

D. Water Scarcity

Freshwater is essential for life itself. It is also essential to a range of activities that support modern civilization, including agriculture, manufacturing, construction, energy production, and sanitation.

Global water supplies are expected to come under greater stress due to increased demand caused by population growth, rising wealth levels, dietary changes, urbanization, climate change, and rising industrial demand. Since most of the water that humans use goes into the production of food, water problems are likely to affect food price and availability.

Currently, around two billion people lack access to safe water for drinking.[60] By 2050, demand for water will have grown by 40 percent, and 25 percent of people will live in countries without sufficient access to clean water.[61]

Climate change is disrupting weather patterns, leading to unpredictable water availability, exacerbating water scarcity, and contaminating water supplies. Perhaps the two most dire impacts of climate change on water are (1) the potential for megadroughts, which could lead both to severe food shortages and to mass migrations toward regions that currently are less vulnerable,[62] and (2) the dramatic loss of snowpack and glaciers, which currently provide water to two billion people globally.[63]

E. Chemical Pollution

The pesticides most widely used currently, collectively known as *neonicotinoids*, are responsible for widespread crashes in the populations of bees and other insects, and for damage to aquatic ecosystems worldwide.[64]

Fertilizer runoff from modern farming acts as a nutrient to algae in lakes, rivers, and oceans, which proliferate and then sink and decompose in the water.[65] The decomposition process consumes oxygen, depleting the amount available to fish and other aquatic life.

Air pollution—including fine particulate matter and toxic chemicals—from burning fossil fuels accounts for an astonishing one in every five deaths worldwide.[66] The pollutants from burning coal in China alone may be causing hundreds of thousands of premature deaths per year,[67] with millions more facing shortened lives. A combination of firewood, biomass, and coal burning has similarly resulted in deadly and worsening air quality in India.

Plastic pollution has formed giant floating gyres in the oceans, and it's been projected that by 2050 the amount of plastic in the oceans will outweigh all the remaining fish.[68] What's more, plastic packaging leaches small amounts of organic chemicals, some known to cause cancer, into the foods they're meant to protect.[69] Many of these chemicals mimic the action of hormones, and are believed to contribute to rising rates of diabetes, obesity, and fertility problems both in humans and other animals. If human sperm counts continue falling at the current rate, they could reach zero (on average) as soon as 2045.[70]

Polyfluoroalkyl substances (PFAS) are one hormone-mimicking class of chemicals used, for example, in making non-stick cookware, stain-resistant carpet, and firefighting foam. These chemicals are now widespread globally and have been linked to multiple health problems, and some can persist and accumulate in the body for years; studies have estimated that PFAS are present in the blood of most Americans.[71]

F. Resource Depletion

Humanity uses two broad categories of resources: renewable and nonrenewable. Some renewable resources are practically inexhaustible (e.g., wind, sunlight, geological heat). Others, however, can only replenish themselves over time (e.g., forests and fisheries). If replenishing resources are harvested faster than their natural recovery rate they will be depleted. Unfortunately, humanity's appetite for critical renewable resources is outstripping their ability to replenish themselves. For example, global fish stocks and primary forest cover have been declining for decades.[72]

Nonrenewable resources (minerals and fossil fuels) do not replenish themselves, thus their supplies are finite. Some minerals and some materials made out of fossil fuels (mainly plastics) can be recycled, but few currently are. Since nonrenewable resources are typically harvested using the low-hanging-fruit (or "best-first") principle, the most common symptom of

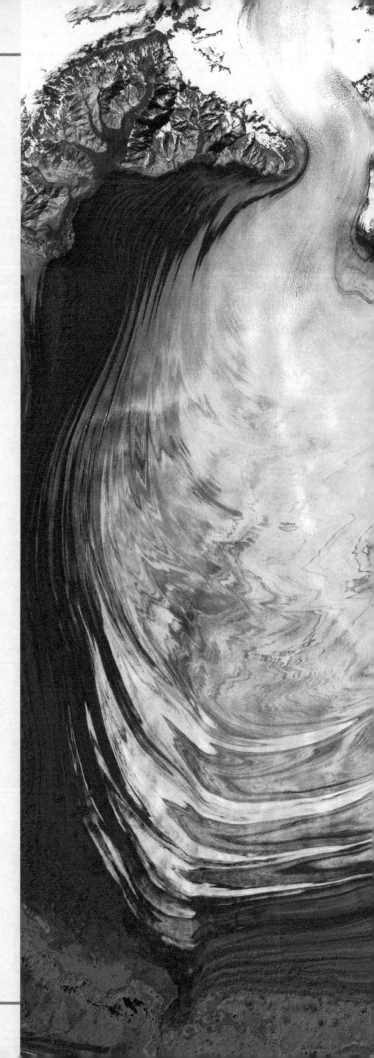

depletion is declining resource quality. As just one example, average copper ore grade has decreased by approximately 25 percent in just ten years[73], while the energy used in copper mining and the amount of waste produced has increased.

Some of the nonrenewable materials needed for the renewable energy transition may become scarce during the current century. According to a report by the IEA, global annual extraction of lithium is expected to increase to 42 times current rates by 2040, graphite to 25 times, cobalt to 21 times, nickel to 19 times, and rare earth minerals to 7 times.[74] The report suggests that available resources are sufficient for a one-time buildout of renewable energy infrastructure to replace current fossil fuel consumption, though some bottlenecks are identified. However, it's unclear whether there would be enough resources to replace that first generation of renewable energy infrastructure once it begins to deteriorate, let alone subsequent generations.

Sand is also likely to become scarce. Desert sand (in which grains have been rounded by friction) is essentially useless in making concrete, which is employed in nearly all construction projects; instead, sand recovered from certain beaches or mined from particular locations (in which individual grains are angular) is needed. In addition, high-purity silica sand used in making glass and semiconductors is depleting at alarming rates.[75]

Some materials are used in ways that make them hard to recycle (such as the trace amounts of gold and other rare metals used in circuit boards). Also, many materials degrade as they are repeatedly recycled.

G. Summary: Feedbacks and Underlying Drivers

It is misleading to think of the problems discussed above as isolated glitches in an otherwise functional and sustainable system of humans interacting with nature. They are, rather, indications of systemic failure.

Over the last 70 years, fossil fuels have enabled massive population growth and consumption growth. As growth has become institutionalized in industrial societies, the process of expansion has become a self-reinforcing feedback loop. Increased energy usage enables population and consumption growth, which in turn incentivize more energy usage. The continuous growth embodied in this feedback loop has spurred climate change and other forms of pollution, while intensifying resource depletion.

Further, there are some less obvious feedbacks among the problems themselves. Water scarcity (worsened by climate change) has led some communities to adopt desalination technologies—which use more energy than conventional water delivery systems, thereby increasing greenhouse gas emissions. The depletion of easy-to-extract minerals leads extractive industries to target lower-quality ores, requiring more energy and producing more polluting waste, often leading to the degradation of waterways and aquifers, thereby contributing to the scarcity of clean water.

There are, of course, potential policy responses to each of the problems discussed above. However, most of those potential responses are inadequate to address the full scope of the problems, are unscalable, or threaten vested interests.

Consider just the climate crisis. The obvious solution is simply to transition from fossil fuels to renewable energy sources. However, despite the fact that this solution has been available (at least in principle) for decades, it has not been widely adopted. Why? The deployment of renewable energy sources at scale faces barriers of cost, materials, engineering, and especially political challenges.[76] Key technologies in sectors such as agriculture, aviation, trucking, shipping, and manufacturing have been designed to run on fossil fuels and will be difficult to electrify (so as to be powered by renewable energy). Concurrently, full climate change mitigation would require transformations in land use including dramatic reductions in cattle raising and deforestation, along with large-scale biosphere restoration efforts. Altogether, what is required for minimizing global warming is a nearly complete redesign of infrastructure and land use patterns. At every step, efforts along these lines must push against entrenched economic and political interests (often and especially fossil fuel interests). But if such efforts fail, the only mitigation responses left are capturing and storing carbon dioxide or geoengineering (i.e., solar radiation management). The former comes with enormous energy, material, and financial costs; the latter carries enormous risks; and the viability of either is questionable at scale.

Similar institutional, economic, and technical roadblocks have stalled solutions to the crises of water scarcity, soil degradation, pollution, biodiversity loss, and resource depletion.

If policy makers had seriously tried to avert environmental unraveling, they would have sought to address its underlying drivers (economic growth, population growth, and unsustainable rates of energy usage and resource consumption), and would have strengthened balancing feedbacks (for example, by reforming industrial agriculture to build topsoil instead of depleting it). Lacking this systemic intervention, what's left is a plethora of targeted, small-scale policy efforts to address the very worst symptoms of impending crisis—attempts that are ultimately destined to fail, since the economic processes that drive crisis upon crisis are still in place.

III. Social Unraveling

Dealing deliberately with the worsening environmental predicament described above will require humanity to drastically reduce its consumption of resources, with most cuts needing to come from wealthy countries and the wealthier sectors of poorer nations. Cooperative effort and, in many cases, shared sacrifice will be needed if social cohesion is to be maintained while such cuts are made. But there are signs that, in many societies, cohesion is already strained.

Those who were exploited or left behind in the march of industrialization, as well as young people in most societies[77], are coming to realize that they will likely never experience the security and abundance that characterized life in the middle-to-upper global wealth strata during most of the past century.

Meanwhile, psychological research shows that those with social power tend to lash out when their power is threatened, often scapegoating the less powerful.[78] This behavioral tendency may feed efforts to increase ethnic and religious distrust in order to splinter popular consensus for progressive economic policies. Extreme political tension is likely to grow in the context of a highly unequal world wherein collective survival may depend on those with the most power (including middle classes in relatively wealthy countries) surrendering some of their advantages.

The following social trends (poverty, inequality, racism and other forms of discrimination, etc.), which derive from economic, political, and cultural conditions, are currently more apparent in some countries than others. But, wherever they appear, they make it more difficult for societies to adapt and respond to the challenges ahead.

A. Poverty

In pre-industrial times, most people survived with little material or monetary wealth. However, in subsistence cultures people tended to share what little they had, making life relatively secure and happy except in times of war or famine. The problem of continual, degrading poverty largely emerged through the destruction of traditional cultures by the processes of colonization, privatization, industrialization, and globalization—processes which, at the same time, made others rich.[79]

Today, the people of the Global South, who generate roughly 80 percent of global wealth through their labor and natural resources, receive about 5 percent of the economic benefits.[80] In 2019 (the latest estimates available), 9.3 percent of the world's population lived in extreme poverty—that is, on less than $2.15 (2017 US dollars) a day.[81] Half of the over 600 million people living in extreme poverty globally live in five countries: India, Nigeria, Democratic Republic of the Congo, Ethiopia, and Bangladesh.[82]

While the number of people living in extreme poverty (as defined by income) has fallen in recent years, the statistic is deceptive. Economic inequality has grown (see below), as has the number of people lacking basic access to food, water, and housing. About 13 percent of people globally do not have access to electricity, and 40 percent do not have access to clean fuels for cooking.[83] Malnutrition is the leading cause of poor health and death around the world, with 1 in 9 people hungry or undernourished.[84]

The consequences of climate change are already falling mainly on poor countries, and poorer communities within countries, and this trend is likely to continue and worsen.

B. Inequality

If poverty is a problem, so is too much wealth if it is distributed highly unequally—either within nations or between them. That's because extreme inequality undermines social cohesion and trust (putting aside questions of morality or justice).[85]

According to the UN's World Social Report 2020, inequality is increasing, driven by four "megatrends": technological innovation, climate change, urbanization, and international migration.[86] The report shows that income inequality has risen in recent decades within most wealthy countries, as well as some middle- and lower-income countries (notably China and India, which together account for nearly a third of the world's population). Among industrial nations, the United States is by far the most top-heavy, with much greater shares of national wealth and income going to the richest 1 percent than any other country.[87]

Economic inequality *between* countries has declined during the same period,[88] but the data are skewed by the inclusion of China and (to a lesser degree) India, whose large economies have grown rapidly. Disregarding these two countries, the income and wealth gap between rich versus poor countries has declined very little. Indeed, the absolute gap between the mean per capita incomes of high- and low-income countries has substantially increased.[89]

Between 1990 and 2015, the richest one percent of global population increased their share of income substantially. Today, people with over $100,000 in assets, who make up 13 percent of the global population, own 85.6 percent of overall wealth.[90]

The COVID-19 pandemic worsened global inequality. Not only did high-income nations receive a far higher share of vaccine doses than poorer nations, but during the pandemic global billionaire wealth increased by roughly $5 trillion,[91] while global workers' combined earnings fell by almost as much, as millions of people lost their jobs.[92]

C. Racism and Other Forms of Discrimination

While there is little evidence in human genome research that race is a useful concept in the genetic classification of humans,[93] history is full of instances of people exploiting or mistreating one another based on perceived physical, ethnic, religious, or cultural differences.

The concept of race, and racial hierarchies, was used as justification for both European colonization and the Atlantic slave trade, and colonizers from around the world have used it as a collective excuse for further inhumane actions. Some former European colonies—such as the US, South Africa, and Brazil—have brutal histories of racial oppression and discrimination, which started in a calculated process of economic exploitation. Even within Europe itself, there is a horrific precedent of ethnic prejudice contributing to genocide. These legacies have intergenerational reverberations.

In the late 20th century, some nations (including the US) made strides toward the reduction of systemic discrimination based on race or ethnicity. However, the past decade has seen a rise of far-right movements in Europe and North America that espouse white nationalist, anti-immigrant views. White nationalists use higher rates of immigration (particularly from regions that differ ethnically, linguistically, or religiously) as a rallying cry, stoking fears of "dilution" of the dominant culture. They also seek to stall or reverse efforts to acknowledge and redress past injustices. Ethno-religious nationalism and persecution of ethnic minorities has also increased in China, India, Myanmar, Iran, Pakistan, Saudi Arabia, Israel, and other countries.

Causes of rising migration include economic and political turmoil, particularly in Africa, the Middle East, and Latin America.[94] But even larger flows of migrants and refugees are likely in the future due to climate change. Unchecked, racist attitudes among dominant groups could erode social cohesion, heighten systemic inequality, increase political polarization, and lead to needless persecution, or even genocide, of minority groups already suffering privation.

D. Antisocial Responses to Scarcity

Social scientists studying modern human responses to natural disasters or to sudden collective deprivation have noted a typical pattern of behavior: initially, people pull together.[95] They share what they have, volunteering their efforts to help neighbors and strangers. However, if scarcity continues for many months or years, then cooperative behavior gradually dwindles, and each individual's circle of trust diminishes significantly.

Often, scarcity can lead to violence—both within and between societies—though the linkage is usually indirect.[96] For example, in Pakistan, rapid population growth, environmental degradation, and inefficient farming practices caused increasing scarcity of both cropland and water by the early 1990s.[97] A resulting urban influx of migrants altered the ethnic balance in the cities, leading to long-running conflict.

Sometimes, leaders stoke the fires of war with other nations in hopes of obtaining control of scarce resources or simply as a way to maintain domestic cohesion. Archaeologists and historians have noted that earlier societies experienced higher levels of warfare when faced with resource shortages brought about by population growth or persistent drought.[98]

Other times, elites compete among themselves, leading to factional division and civil war. Examples that emerge from historical studies of environmentally-driven civil conflict include the Chiapas rebellion, the Rwandan genocide, violence between Senegal and Mauritania, civil conflict in the Philippines, and ethnic violence in Assam, India.[99] In civil disputes such as these, minorities nearly always suffer the worst casualties.

Foreseeable triggers for future conflict center on the impacts of climate change, population growth, resource scarcity, and environmental degradation. At the same time, in recent decades the means of conflict have proliferated, as

weapons have grown more numerous, deadly, and sophisticated—now including (globally) an estimated one billion guns, 14,000 nuclear warheads, and new cyberweapons capable of crippling power grids or energy and water supplies for entire nations.[100]

E. Authoritarianism

Democracy is a system of governance designed in part to prevent the rise of tyrants. But, for its maintenance, democracy requires trust, cooperation, and a general willingness to follow rules—along with accurate, verifiable, and widely available information about topics that affect citizens' lives.

Many modern democracies are multi-ethnic, multi-religious, and even multi-linguistic. They are held together by the shared belief that their governments, for the most part, are fair and provide equal opportunity and protection. When that belief erodes, so does the legitimacy of government.

Since the 1980s, social scientists have largely agreed that "substantial and persisting increases in the scarcity of widely-sought resources in contemporary societies" lead to "a shift from open toward more closed and authoritarian political institutions."[101] Historic examples include the Great Depression, which brought threats to democracy worldwide from both the left and right ends of the political spectrum.

About half of all countries are currently democracies of some kind. In recent years, the Democracy Index has shown a decline in democratic institutions and engagement.[102] The trend worsened during the COVID-19 pandemic, with almost 70 percent of countries recording a reduction in their overall scores, and the averaged global score hitting an all-time low.

F. Impacts of Technological Change

Technological change nearly always produces winners and losers. Today, as in earlier phases of the industrial revolution, those who control new technologies are likely to be winners—but in the foreseeable future, the losers could number in the billions. The technologies discussed below have great potential to worsen already high economic inequality.

Most futurists anticipate that artificial intelligence (AI) will have significant impacts on current employment patterns, with hundreds of millions of machine and vehicle operators, service workers, support staff, and white collar jobs around the world potentially being rendered jobless. It could also significantly increase global energy demand. Some analysts estimate that a single fully autonomous vehicle would consume as much as five terabytes of data in an hour,[103] roughly the equivalent of an average American household's internet data usage for an entire year. Already, the internet and data processing are responsible for over ten percent of global electricity demand. Under foreseeable AI growth scenarios, that proportion could more than double.

The continuing development (by state intelligence agencies and rogue non-state groups) of computer technologies for purposes of subversion is leading to increased risk of cyberterrorism, including threats to infrastructure such as water systems and electricity grids. Meanwhile, autonomous weapons now under development could provoke a new arms race and lead to an ominous new brand of warfare in which attackers are more shielded from accountability.

As electronic tools become more sophisticated, authoritarian states and private companies can surveil citizens, workers, and consumers ever more thoroughly. Also, as citizens become more dependent on the internet, governments can exert greater social control by restricting internet access to shut down their opposition (as has happened in Tibet, Myanmar, and Russia).

Social media's tendency to amplify users' existing views by algorithmically filtering news and opinion posts is resulting in increasing political polarization, the proliferation of conspiracy theories, and a withering of social trust.[104] Meanwhile, distributed information networks are contributing to the decay of trust in institutional information sources (principally, the news media and academia).

As entire societies become more dependent on computer-mediated communication and supply chains, risks of catastrophic breakdowns multiply. As just one example, a once-in-a-century massive solar flare could fry electronic devices and cripple electricity grids worldwide.[105] When Earth passed through such a flare in 1859, the result was merely an interruption in telegraph communications. If the same thing were to happen today, global chaos might ensue.

G. Summary: Feedbacks and Underlying Drivers

These social and political problems are not isolated from one another, nor are they the result of happenstance; rather, they are related symptoms of systemic failure. At the root of poverty and inequality is the predictable tendency of those who already have an economic advantage to use it to increase their advantage further. It's a tendency that goes back to the origin of money and social hierarchies at the dawn of civilization.

Societies long ago recognized the destabilizing results of this self-reinforcing feedback process, and sought ways to counteract it through cultural taboos against hoarding wealth, debt jubilees, constitutional limits on political power, the establishment of trade unions, and other means. Nevertheless, throughout history, most decisive checks on unequal social power have occurred not by way of social innovation, but as a result of war, revolution, or economic collapse. The wealthy exploit resources and labor until the social and ecological systems can tolerate no further exploitation.

In the present era, fossil fuels have enabled exploitation to become globalized. Socio-ecological limits have thereby been expanded to a planetary scale, and a reckoning was delayed. However, the process of exploiting people and nature is now confronting those expanded limits.

The consequences of this collision with socio-ecological limits are themselves subject to cross-influencing feedbacks. Unequal growth within and between societies eventually leads to a breakdown of trust throughout societies and communities, thereby feeding trends toward authoritarianism, which lead to even more breakdown of trust.

Further, self-reinforcing feedbacks between ecological breakdown and social breakdown are strengthening and growing more numerous. For example, climate-driven human migration presents challenges to political systems while also eroding traditional cultural norms that support environmental stewardship. Societies in the midst of social crisis, and ones turning toward authoritarian regimes, are seldom able to muster efforts toward resource conservation, emissions reduction, and habitat preservation; indeed, under such circumstances, past efforts in these directions may be undermined.

What would it have taken in the past for leaders to have fully averted our current social unraveling? Policy makers would have had to more effectively address its main driver: socio-economic inequality, both within and between nations. They also would have needed to strengthen balancing feedbacks by respecting Indigenous cultures and land rights, while helping people in urban industrial societies to find security and happiness in their local regions and in patterns of life that are low in resource consumption, while high in aesthetic and cultural satisfaction.

IV. Pulling on the Threads: The Interaction of Environmental and Social Drivers

The COVID-19 pandemic has delivered, for many people (particularly those who had previously lived in conditions of relative stability), a painful lesson: none of us is immune to the forces of nature, to events across the globe, to behaviors and choices of others both near and far, or to complex and often invisible systems interconnections—including those embedded in global supply chains and social media platforms.

The pandemic has also served as a test case for how factors driving the Great Unraveling—in this case, increasing inequality, antisocial responses to scarcity, and the influence of technology—can play out in the real world. The politicization of public health responses and the extreme disparity in vaccine availability between the wealthiest nations and poorer nations encompassing the vast majority of the world's people do not bode well for our capacity to respond collectively to the even more dire threats approaching.[106]

A. The New Reality

For many in the United States and elsewhere, recent years have felt like a dizzying whirlwind of environmental and social emergencies: political polarization run amok and democratic institutions under attack; record-breaking heat waves, floods, droughts, and fires; the shattering of consensus on basic facts; the pandemic and the divergence of responses to it by leaders and the general public; and glaring manifestations of racial and economic inequality.

Unfortunately, the general view remains that these are all isolated problems that will be overcome in due course.[107] There is as yet little recognition that these challenges are *systemic*, *entrenched*, and *interrelated*. It's not just the public or policymakers who are failing to see the connections. Even among those working on the frontline of individual environmental or social issues, few seem to recognize their chosen issue within the context of complex interactions between numerous environmental and social systems, all undergoing varying degrees of destabilization, breakdown, or collapse.

It's true that some groups of activists and scholars, including ones in the climate movement, have made laudable steps towards embracing "intersectionality"—including intersections between the climate crisis and racial/economic justice issues. But that intersectionality is often viewed as the interactions between a relatively small number of discrete issues that can be "solved" one by one. The Green New Deal (GND), for example, is championed as a strategy for mitigating the climate crisis and reversing inequality by providing millions of well-paying manufacturing jobs to deploy renewable energy technologies. However, at least among more mainstream GND advocates, this strategy does not incorporate an understanding of many critical drivers—in particular:

- The depth of the role that fossil fuels currently play in all sectors of society;

- The limitations and challenges in scaling and maintaining a renewable energy system that can fully replace fossil fuels and even meet expectations of continuous energy demand growth;

- The transformations that will likely occur in most aspects of modern life as a result of the energy transition; and

- The rippling economic and social impacts of that transition, especially if undertaken at the speed that climate mitigation demands.

Without a full appreciation of this big picture, it will be difficult for GND advocates to plan and implement an effective strategy, and they will not make substantial progress on the problems that so desperately need to be solved.

Climate activists and organizations can be forgiven for this failure. They face relentless pressure from supporters and donors to achieve the Herculean task of eliminating gigatons of global greenhouse gas emissions in a matter of years. They've had to grapple with the complexities and changing understanding of climate science while engaging in the challenging work of movement-building and policymaking. Similar dynamics play out among organizations and institutions that focus on other problems.

The recent widespread adoption of the term "polycrisis" reflects a growing recognition that we must contend with multiple challenges simultaneously;[108] however, a deep analysis of causal links and dynamics between these

Sidebar: The Spectrum of Destabilization > Breakdown > Collapse

When we refer to the condition of various systems or the entire state of society, terminology becomes challenging for two main reasons: 1) it is difficult to empirically and consistently quantify conditions across distinct systems like global finance or a local coral reef, and 2) terms like "collapse" tend to engender subjective responses. As resilience scientists Graeme Cumming and Garry Peterson have observed,

> "…the questions of how much and what kind of change constitutes a collapse, whether fast and slow changes both qualify as 'collapse', and whether collapse must have a normative dimension (and if so, then who decides on that dimension, since it may depend on perspective) are all contested." [113]

Cumming, Peterson, and many others have attempted to correct for often arbitrary and conflicting uses of concepts like "collapse" within academic literature by developing more comprehensive frameworks. While such frameworks are valuable, our intent here is simply to provide those concerned with systemic environmental and social challenges with clear and consistent use of terms like destabilization, breakdown, and collapse—particularly as they relate to one another on a continuum. The far end of that continuum is collapse. For our purposes, the collapse of a system—whether a pine forest, the market for insulin, or the global economy—entails a lasting loss of functioning or the loss of the entire identity of that system, as when a savannah ecosystem shifts to a desert.

The difference between the destabilization of a system and its breakdown can be more difficult to define, as it largely pertains to degree of severity. When a system undergoes destabilization, it experiences notable changes in its functioning or behavior. This destabilization is not merely an isolated event or perturbation, as when the Dow Jones Industrial Average (DJIA) experiences a dramatic, short-lived loss in value, but rather a pattern of repeated discontinuities. These patterns of discontinuity could last a long time—for example, if the Dow Jones undergoes repeated, dramatic jumps and drops in value for many months—to the point where such patterns are viewed as "the new normal." What's key is that these patterns reflect instability. What differentiates breakdown from destabilization is that a system undergoing breakdown experiences a profound loss of function or structure, though this is neither permanent nor entails the loss of that system's identity (that would be its collapse).

Sticking with the example of the Dow Jones Industrial Average, its destabilization could entail a pattern of dramatic, unpredictable shifts in value (up 8 percent one week, down 5 percent the next, up 3 percent again, down 9 percent again, and so on for a significant period of time); its breakdown would occur when all that instability leads to shareholders suddenly fleeing the market in large numbers, compelling the DJIA operators to temporarily suspend all trading); and its collapse would stem from such a loss of value or number of traders that a critical mass of companies listed in the market abandons it, or it is forced to close permanently.

various crises is still much needed.[109] The persistent inability to integrate both a systemic understanding of this confluence of complex environmental and social crises *and* the likelihood of an accelerating unraveling of systems increases the odds that our efforts to eliminate, mitigate, or adapt to specific crises will fail, and that key systems—indeed, society as a whole—will tip from destabilization to collapse.

B. Destabilization, Feedbacks, and Conclusions

Unfortunately, there is no shortage of examples of destabilization already under way, and destabilized systems often possess the ability to spread destabilization to other systems. A simple example of mutual destabilization can be seen in the effect of climate change on human migration. The burning of fossil fuels releases vast amounts of greenhouse gas emissions, which raises global temperatures. This rise in temperature leads to increased frequency and severity of wildfires, flooding, droughts, and other natural disasters. These acute and chronic disasters lead to the displacement of a growing number of people. In this case, a fairly simple progression of destabilization can be recognized: fossil fuel burning → climate change → natural disasters → human migration.

It is more challenging to recognize complex and interactive dynamics across multiple systems, and amplifying feedback loops within and between systems. Thinking about these system dynamics helps us better understand the real world. For example, to continue with our example of interactions between climate change and human migration:

- How might the nature of the climate crisis, including its speed, severity, and localized characteristics, influence the movement of people?

- What happens to communities or societies that experience significant immigration of people displaced by climate change?

- How much does the rate of immigration impact the level of destabilization of those communities?

- How do the cultural characteristics of incoming migrants affect their integration with existing residents of a community?

- How is local culture affected when a community experiences significant out-migration?

- What are the economic impacts, e.g., the cost of housing or wages and employment?

- What are the environmental impacts, e.g., on local ecosystem health, water resources, and farmland productivity?

- What are the political ramifications locally and more broadly?

- And, ultimately, how do these various dynamics—the rate of change, human psychology, culture and collective behavior, environmental and economic impacts, and politics—interact to determine whether climate-driven migration leads to destabilization, breakdown, or collapse of the communities impacted?

These questions are not just of hypothetical interest for forward-thinking policymakers and planners. They have profound implications for climate activists; for anyone working to protect human rights, economic and racial equality, and the stabilization and spread of democracy; and indeed, for anyone trying to navigate and mitigate cascading crises.

Unfortunately, trying to accurately model, let alone predict the outcomes of, interactions between multiple systems is a task that inevitably comes with significant uncertainty. And yet we are far worse off if we ignore these complex interactions. At the least, we must collectively recognize and come to terms with some difficult facts, which build upon one another:

- **We now inhabit a single socio-ecological system, or metasystem.** The scale of human activity across the planet, coupled with the connectivity and interdependence of the global economy, means that destabilization, breakdown, or collapse in one area is likely to have ripple effects throughout the metasystem. The tight coupling of economic, energy, and environmental systems increases the risk of cascading crises or synchronous failure, should just one part of a system fail.[110]

- **Destabilization is already upon us.** As summarized in Sections II and III above, many environmental and social systems are already experiencing accelerating destabilization. Ironically, taking aggressive action to minimize the risks of breakdown or collapse—for example, slashing global emissions by 45 percent by 2030 as recommended by the Intergovernmental Panel on Climate Change—would itself be highly destabilizing.[111]

- **Our delay has made the challenge much bigger.** Since 1980, global GDP has grown 230 percent, human population 78 percent, energy consumption 112 percent, and greenhouse gas emissions 66 percent.[112] Some pundits argue that this growth increases human capacity for problem solving. However, an understanding of the dynamics of destabilization tends to lead to the opposite conclusion: that much greater efforts are needed now to alter human systems that are even more deeply entrenched and that are much closer to destabilization.

- **Adaptation will be exceedingly difficult.** Systemic interconnections, unfolding destabilization, and the need for profound transformations in how we conduct daily life (whether by choice or by force of circumstance) mean that the new status quo, if we can call it that, will consist of tremendous change and upheaval. Instead of a "new normal," we must adapt to situations in which the word *normal* no longer has meaning.

V. Weaving a New Tapestry: How to Respond to the Great Unraveling

Faced, as we are, with the rapid unraveling of the tapestry of our environmental and social systems, it is hard to imagine doing anything but trying our best to stitch that fabric back together again. It's natural to want the world we know to continue. However, the picture looks different depending on one's vantage point. For people in impoverished nations or communities, the status quo has long consisted of political upheaval, economic hardship, broken social systems, and a degraded local environment. Still, nearly all societies have enjoyed a protracted period of global climate stability (the Holocene), and, during recent decades, have adopted the neoliberal economic vision of growth and progress, as well as the globalization of the economy. The background assumption has been that our current way of life can be maintained and improved.

If we are, in fact, unable to stitch this tapestry back together, what should we do? Perhaps the hardest part of any process that entails making significant changes and establishing new behaviors is the beginning. To mount a fitting response to the Great Unraveling, we must begin from within ourselves. We have to learn how to navigate uncertain conditions and uncertain times. And then we have to work together to apply our learning and skills to the scale of our communities, and then the scale of our nations and the planet as a whole.

A. Navigating the Liminal Space

Most modern societies have been steeped for generations in some version of the myth of progress—the expectation of a future with more economic and educational opportunities, more political stability, more technological advancement, and more peaceful global political integration. This myth of progress saturates nearly all levels of society, from parents' hopes for their children's futures to the assumptions made by policymakers, educators, owners of capital, and business leaders. When prompted to picture a less rosy future, many can only imagine a Hollywood-style cataclysmic event—a nuclear holocaust, alien invasion, or zombie apocalypse.

The Great Unraveling challenges us to grapple with the prospect of a more complicated and nuanced future, one of compounding crises— some acute, others chronic—interacting across environmental and social systems in complex ways, at different rates, in different places, and with different results. Critically, unless a global apocalyptic event does usher in a sudden and complete collapse of society, humanity will still be here and will have some amount of agency over how the unraveling unfolds and what comes after. Our story does not end when global temperatures reach 1.6°C or 2.1°C above pre-industrial levels, when the global economy falls into a depression, or even when our communities have been ravaged by a natural disaster or violence.

Stories of progress and apocalypse are diametrically opposed, but they provide a similar psychological release from uncertainty. Faith in progress, or fatalism about humanity's march towards extinction, are two sides of the same coin, a coin that affords the bearer a reprieve from reckoning with the reality we will likely face—a reality that is at once unspeakably challenging and pregnant with possibility and responsibility.

Holding this space between—*the liminality*— requires us to think the unthinkable, to accept uncertainty, to resist both hopelessness and blind optimism, to stretch ourselves personally and professionally, and to practice self- and collective care. What's coming will affect each of us differently, and will likely pose constant challenges, as we contend with the psychological desire to alleviate dissonance; the practical, social, and institutional pressures of everyday life; and the unfolding of the Great Unraveling itself.

B. Recognizing Obstacles to Systemic Change

Finding the agency and means to navigate the unraveling of environmental and social systems, and to prevent their collapse, requires a recognition of some of the obstacles keeping us from intervening in meaningful ways. Four main categories of these obstacles are:

1. Biophysical realities and the various constraints they impose;

2. Cognitive bias and related glitches in individual and collective behavior;

3. Entrenched socioeconomic systems and belief systems; and

4. Diminished capacity to act systemically or pro-socially.

Biophysical Realities

Ultimately, how we navigate the unraveling of environmental and social systems comes down to human agency, choice, and leadership. However, human agency is constrained by physical reality—most relevantly, the viability and health of ecosystems, the availability and quality of natural resources and energy sources, and the capacity of the planet to absorb wastes and support human activities.

We have already discussed environmental challenges (climate change, biodiversity loss, fresh water scarcity, resource depletion, and chemical pollution) in Section II above. Humanity can reduce the severity of these environmental challenges or the risk of crossing calamitous thresholds by massively and rapidly reducing habitat loss, pollution, and material consumption. Making these sorts of changes might entail, for example, building more public transit infrastructure to reduce societies' reliance on inherently inefficient automobile transport. But making significant changes to infrastructure requires energy, materials, and access to natural resources (including land, water, and ecosystem services). We've already depleted, polluted, or overused many of these resources: aquifers have been drawn down, high-quality ores are dwindling, and forests have been logged.

Most crucially, replacing our global fossil fuel-based energy system with alternative low-carbon sources requires enormous amounts of energy for building solar panels, wind turbines, batteries, transmission lines, and vehicles and equipment that run on electricity.[114] Altogether, this will amount to the largest industrial project ever attempted—by far. In the early stages, most of energy for this undertaking will have to come from fossil fuels, since coal, oil, and natural gas supply about 85 percent of all global energy currently. Therefore, the energy transition will itself cause a significant pulse of carbon emissions. Yet our remaining carbon budget (the carbon that humanity can emit without triggering 1.5° or 2°C of warming) is already down to just a few years of current business-as-usual emissions.

So, ironically, the energy transition—which we are undertaking to reduce carbon emissions—may end up causing humanity to blow past the emissions targets commensurate with maintaining a livable planet. The only way to avoid this conundrum would be for industrial nations to greatly reduce energy usage for "normal" operations—most especially transportation and manufacturing—as soon as possible. But this would amount to deliberately shrinking the world's largest economies. Would that be politically feasible? What would be the consequences for the global economy, and for the lives of billions of people?

The undeniable reality of biophysical limits forces us to confront profound questions like these:

- **Can we slash humanity's planetary footprint while still growing the global economy?** Current analysis suggests the answer is no.[115]

- **How many humans can the planet sustainably support and at what level of *equitable* material consumption?** Estimates vary greatly, but conservative ones that assume a generally shared industrial level of production and consumption trend toward a global population limit of one to three billion.[116]

- **To what degree can renewable energy technologies replace fossil fuels sustainably, equitably, and with limited environmental and human impacts?** Again, estimates vary greatly, but the most optimistic projections tend to rely on unrealistic assumptions.[117]

- **What impact could the declining availability of fossil fuels and other critical nonrenewable resources have on the energy transition and on society more broadly?** According to at least one systems dynamics analysis, if supplies of fossil fuels and critical minerals are constrained, the energy transition may not be physically achievable without radically reducing society's overall energy and material throughput.[118]

- **How much should we rely on negative emissions technologies (heavily factored into climate stabilization models) to remove carbon dioxide from the atmosphere?** The best study so far on the prospects for carbon capture concludes that only biological methods (such as capturing CO_2 in forests and soils) are actually viable at scale.[119]

Individual and Collective Behavior

Human psychology and evolutionary biology profoundly affect how we view, interpret, and respond to environmental and social crises. Though largely invisible and often irrational, our psychological and biological imperatives promote deeply ingrained behaviors that present a formidable obstacle to achieving systemic change. These behaviors are adaptive traits that

helped Homo sapiens survive for 99 percent of our history as a species. Unfortunately, many of these traits are maladaptive for managing the global industrial society we've built over recent decades and the crises that this society has engendered.

A few of the many cognitive biases[120] and other behavioral traits that make it challenging to address environmental and social crises include:

- **Discounting the future.** For most of human history, it made sense to favor present rewards over potential future costs, or to resist making sacrifices today for the possibility of future benefits (though some Indigenous societies notably made decisions with future generations in mind[121]). Unfortunately, the existential risks we face today require the opposite response: we must sacrifice now if our grandchildren are to have the opportunity to thrive.

- **Tribalism and in-group/out-group dynamics.** For most of human history, conformity to the expectations of our tribe and distrust of people outside our tribe were keys to survival. Despite the fact that modern economies tend to promote individuality and self-interest, deep-seated in-group/out-group social and psychological dynamics still hold sway. Irrational tribalism—especially in periods of uncertainty or crisis—makes it challenging for individuals to break away from dominant unsustainable or unjust social norms. Tribalism also fuels polarization, divisiveness, scapegoating, and even violence.

- **Confirmation bias and motivated reasoning.** All of us are susceptible to confirmation bias (the seeking out or remembering of information that reinforces preexisting beliefs) and motivated reasoning (the tendency to readily accept things that we want to believe and to be skeptical of things that we don't).[122] These two biases are especially strong when it comes to deeply held beliefs or highly emotional subjects. When coupled with in/out-group dynamics and algorithmically driven sourcing of content online, these cognitive biases make it challenging to think critically or find widespread agreement on the causes of, and responses to, our most pressing challenges.

- **Sunk cost bias.** The bias toward systems, policies, or technologies that have received prior investment often favors unsustainable or unjust solutions. Individuals act with a sunk cost bias regularly—from finishing a book they don't like, because they've already read half of it, to getting a job as a lawyer despite hating the actual work, because they've committed years of their life and tens of thousands of dollars to getting a law degree. Organizations, communities, and nations are also susceptible to the sunk cost bias, and it regularly affects large-scale policy choices, like spending billions of dollars every year to maintain an unsustainable highway system rather than supporting the land use changes that minimize the need for low-density transportation.

- **Patterning.** While the ability to recognize patterns in nature (for example, the behavior of prey animals) was a highly successful adaptive trait in pre-industrial times, the human brain's tendency to seek patterns can drive people to oversimplify or misinterpret situations, and can even fuel belief in conspiracy theories.

- **Novelty seeking.** Novelty seeking, too, is a behavioral trait that likely had adaptive advantages in spurring humans to take risks or pursue new knowledge or skills that sometimes led to big rewards. But our tendency to seek experiences that trigger the release of dopamine (a neurotransmitter associated with intense feelings of reward[123]) can easily be "hijacked," for instance by social media. Moment-by-moment reward seeking tends to shrink our attention and makes it more difficult to prioritize problems that may take generations to solve.

- **Difficulty comprehending the current size and scale of the human project.** Human impacts have outgrown our ability to comprehend, let alone manage them. Some phenomena, like global warming—described by the philosopher Timothy Morton as "hyperobjects"[124]— operate at scales of time and space that defy our ability to grasp them. At the same time, the dynamic of compounded growth—which characterizes everything from global economic activity and energy demand to the amount of internet data being created every hour—is foreign to the patterns of nature and human activity in which we evolved.[125]

Socioeconomic Structures, Institutions, and Belief Systems

Anthropologist Marvin Harris hypothesized that all human societies operate in three realms: *infrastructure* (which consists of our interactions with nature to obtain food and materials), *structure* (the means by which we make collective decisions and allocate resources and wealth), and *superstructure* (the realm of ideas, values, beliefs, and worldviews).[126] Harris argued that cultural change can proceed within each of those realms, but that major infrastructural change tends to drive significant shifts also within society's structure and superstructure. Thus, when certain human societies abandoned the infrastructure of hunting and gathering and adopted the infrastructure of large-scale grain agriculture, their structures and superstructures followed suit. These societies transformed from small, usually egalitarian, animistic bands to nation states with rigid hierarchies of power and religious belief in sky gods.

Viewing modern society through the lens of Harris's cultural materialism, we can see that humanity's growing ecological footprint, the structures of the global economy, and broadly held beliefs in human exceptionalism and unending technological and material progress stem largely from a recent and extraordinary infrastructural shift—the explosion of fossil energy production and consumption, which simply can't persist for reasons explored above. All of the wonders of the "developed" world that influence how we think about the world arose from exploitation of fossil fuels. To avoid widespread collapse of environmental and social systems, humanity now faces an unprecedented challenge—to transform our socioeconomic and political systems (structure) and our belief systems (superstructure) *before* another infrastructural shift (climate change and the depletion of fossil fuels) forces such changes upon us. The point is to get ahead of the curve—to manage the required downshifting and avoid suffering. But to do so, we'll have to invert the usual order in which structure and superstructure evolve.

Perhaps most significant, and most challenging to address are the structures that support the modern economy. The modern global economy is built on large-scale energy usage and material consumption, growth, and exploitation of labor and the natural world. Despite claims that energy and materials consumption, as well as greenhouse gas emissions, are being decoupled from economic growth, the correlation of global energy usage to economic activity remains almost perfect.[127] And continued growth of the economy is not just an aspiration, but an absolute requirement of today's economic and financial playbook, including the way money gets created. Most people think that money is created by governments. But, in fact, banks create most money by loaning it into existence with the expectation of a return of both principal and interest. When a bank makes a loan, it records the amount as an asset in its own account, and a liability in the debtor's account. The money itself appears at the moment the loan is recorded. Without economic growth, interest cannot be paid on most loans, and a large-scale round of defaults will ensue, wiping out most investors and much of the financial system itself.

Growth is also an expectation of virtually every investor-backed business (whether publicly-traded or venture capital-backed) and exists as a largely unquestioned foundation of everything from tax-based government to climate models to the operation of philanthropic institutions.

But growth is unsustainable on a finite planet and requires the exploitation of nature and people. Many pundits promote future scenarios where economic activity continues to grow but we somehow avoid exploitation of people and planet. Such scenarios almost universally depend on technological solutions—negative emissions technologies, renewable energy, and artificial intelligence—that don't exist, can't support growing energy demand, or require vast increases in the use of energy and material resources.

Our dependency on growth also has psychological and cultural dimensions. For decades, economic growth has been equated with "progress"—a pairing supported by advertising, politics, and policy. Thus many, if not most, people tend to see giving up growth as accepting a future of decline and failure.

A few other socioeconomic and cultural impediments to the transformative change required to avert the breakdown or collapse of environmental and social systems include:

- **Prioritizing the near term.** Discounting the future is not just an individual cognitive bias; it is baked into economic and political decision-making. In democracies where elections are held every two to six years, there is little incentive for politicians to enact policies to address existential threats like climate change that require near-term sacrifice, cost, or instability. The same dynamic plays out in finance, where short-term decision making (e.g., to achieve quarterly projections) drives the behavior of corporations that control much of the global economy.

- **Complexity and specialization.** We have built a global economy so complex that it requires specialization in all sectors of society. This level of complexity and specialization presents a challenge for any individual to understand the big picture of the global economy. And it's even more challenging to identify and act upon leverage points to transform it.

- **Inertia of the status quo.** People tend to resist change and act on what they know. The familiarity of the status quo (even if it is trending toward unraveling) is often seen as preferable to the uncertainty of change (even if it may come with a major upside). Coupled with sunk cost bias, this preference for the status quo hinders engagement in bold, rapid alternatives to change familiar structures in the built environment and entrenched governmental processes.

- **Belief in individuality and human exceptionalism.** Central tenets of modern, capitalist societies—best exemplified by the United States but shared by much of the world—are the beliefs that pursuing individual self-interest benefits society as a whole, and that humans have the unique capacity to overcome all natural limits through our ingenuity. These beliefs are extremely recent in the scheme of human history and run counter both to deep-seated evolutionary traits and to wisdom gained by Indigenous societies over many millennia. But they serve as a largely unquestioned obstacle to the kinds of

cooperative behavioral and structural changes required to avert collapse.

Diminished Capacity to Act Systemically or Pro-socially

Neoliberal economist Milton Friedman famously stated that "Only a crisis—actual or perceived—produces real change."[128] While it's true that disasters have led to profound shifts in public awareness and have been used to advance otherwise unpopular policies, *positive* transformative change less often follows from disasters.[129] A brief scan of human history sadly offers many examples of crises leading to the rise of autocracies, violence and wars, and persecution or even genocide.

While there is risk of regressive responses to crises, there is also a potential for achieving little to no meaningful change. After all, many existing systems (national governments, capitalist economies, shared belief structures) are highly resilient and have already weathered considerable turmoil. But if these human systems do not change, they will continue to create conditions that give rise to societal breakdown or collapse. A major driver of complacency is the power wielded by those who benefit from the status quo. But lack of change may also simply be the product of diminished capacity to enact reforms—financial, material, or sociopolitical—as destabilization accelerates.

At the same time, the likely rise in the number and severity of acute crises—natural disasters, epidemics, economic contractions, supply chain disruptions, sectarian or ethnic conflicts, and so on—increases the risk that we, as individuals and as whole societies, will bounce from one crisis to another, operating only in reactive mode. Picture how difficult it would be for a fire brigade to work on improving its leadership structure if it's constantly running around town putting out fire after fire.

Any constructive response to the systemic environmental and social crises we face, and to the biophysical, behavioral, and structural obstacles to dealing with those crises, outlined immediately above, must begin with a recognition of both the crises and the obstacles. That's a main goal of this document. But this recognition, though essential, is only a beginning point. We very quickly need to become experts at prioritizing the crises, recognizing their root causes, and overcoming the obstacles to taking effective action. After that must come the hard work of building or rebuilding social and technological systems to the degree we can, dealing with the consequences of systemic change and breakdown, and, to the extent that's possible, preventing harm to people and nature.

C. Getting It Right or Getting It Wrong

The nature of the Great Unraveling—converging crises in numerous complex, adaptive environmental and social systems at scales from the local to the global—makes the task of predicting the future in any detail a fool's errand. But detailed prediction isn't necessary in order to consider what factors and actions are more likely to lead to a more resilient, equitable, and sustainable society versus those likely to hasten breakdown and collapse. What factors and actions might lead us to get our response to the Great Unraveling wrong, or right?

Getting It Wrong

Acting as though the Great Unraveling is not already upon us, and does not require transformative change to navigate it and reduce the risk of breakdown or collapse, is effectively to get our response wrong. That's because acting this way will lead to:

- **Delaying bold action.** The longer it takes to mobilize widespread, ambitious action to address the environmental and social crises we face, the more severe the Great Unraveling will be and the higher the risk of breakdown (or even collapse) of essential systems will be. Delaying action through incrementalism can be nearly as damaging as no action at all. Tackling the issues culminating in the Great Unraveling will require major and rapid directional change, and reallocation of resources, throughout industrial societies. For example, incentivizing the adoption of electric vehicles is an insufficient response to the challenges that climate change and resource depletion pose to our current global transport system. Given limits to materials and energy, our ultimate goal must be an overall reduction in mobility, especially in transport modes such as aviation and personal motorized vehicles—with a relocalization of economies to reduce the distances people and materials need to travel. Moreover, while we pursue this goal we must also take steps to ensure that the essential services (such as food distribution, healthcare, and education) currently provided by highly resource-intensive transport systems are not severely curtailed.

- **Attempting to maintain maladaptive aspects of the status quo.** Attempting to address the systemic risks we face while holding onto the very structures, beliefs, and behaviors that created those risks in the first place is a recipe for disaster. It seems inconceivable that we could transform the global economy away from its fixation on growth, consumption, immediate returns, and efficiency; address disparities in power and wealth; move from fetishizing individual freedoms to nurturing collective wellbeing; or shift from a parasitic relationship with the natural world to a symbiotic one—that is, until we realize the status quo is actually a dangerous aberration in the human story and is leading us toward ever greater destabilization.

- **Using linear and siloed thinking.** Linear thinkers view a problem as a process with a starting point and a sequence of connected steps, ultimately leading to a solution. Simple, isolated problems can often be effectively managed using this kind of thinking. However, when problems are not isolated but interact across technological, social, and environmental systems, linear thinking can then obscure the real situation and lead to outcomes that are unsatisfactory or even catastrophic. Siloed thinkers tend to stay within a strictly defined range of expertise (such as poverty elimination, environmental pollution abatement, or energy management) and have difficulty seeing how developments within their field are tied to those studied by different sets of specialists.

If we continue viewing the crises described in Sections II and III above as isolated problems that can be prioritized and addressed separately rather than systemically, we risk widespread destabilization or cascading breakdowns, even if we have made progress in tackling some individual issues. Likewise, anticipating that destabilization in environmental or social systems will continue linearly (because it has done so to date) ignores the very real threat of sudden and dramatic shifts. The Arab Spring in 2011 serves as an illustration of both of these dynamics at play: chronic conditions of corruption and economic stagnation in the Middle East/North Africa (MENA) region combined with a spike in global grain prices as a result of climate change-influenced droughts created explosive political upheaval throughout the region.[130]

Getting it wrong in our minds will inevitably lead to getting it wrong in terms of behavior and policy. The results, then, are likely to include growing polarization, conflict, violence, inequality, and injustice; the rise of "Green Fortresses" and sacrifice zones; and loss of life, diversity, and possibility.

Getting It Right

In contrast, getting our response right will likely entail:

- **Acting swiftly and aggressively to minimize the risk or extent of destabilization, breakdown, and collapse.** What's needed is a rapid reduction in the brittleness of key human and ecological systems. In our efforts to reduce that brittleness and build resilience, we must start with the systems everyone depends on: food, energy, water, healthcare, communication, and money. There are already efforts underway in communities around the world to reform or transform each of those systems—including efforts to localize food systems, incentivize energy conservation, deploy renewable energy sources, conserve fresh water, improve access to healthcare, reduce the intrusive impacts of modern communication technologies, and create alternative means of exchange. However, such efforts currently operate at an insufficient scale. Policy makers must bring these efforts into the foreground and support them with adequate funding and appropriate policies.

- **Resisting the forces that increase the likelihood and severity of collapse.** Despite clear indications that our current path is ruinous to the wellbeing of billions of people and trillions of non-humans, the status quo of corporate profit maximization, deepening economic inequality, and investments in new fossil fuel infrastructure continues apace. We must strengthen efforts that resist the systems, institutions, policies, and projects which hasten and worsen the unraveling of environmental and social systems. These efforts range from political engagement and shareholder/consumer actions to civil disobedience and direct action aimed at defending communities and ecosystems from extractive capitalism.

- **Choosing the right systems to fortify.** Effort expended in increasing the resilience of systems and infrastructure that are inherently unsustainable is essentially wasted. For example, airports and highways are transportation infrastructure that is inherently inefficient and needs to be de-emphasized if we hope to minimize climate change and other environmental stressors. It is *the essential functions of systems*—the meeting of human needs for food, water, communication, and health—that need to be made more resilient, not the existing means of meeting those needs. The available time and human capacity for building and repairing effective, sustainable systems is limited. Therefore, we will need to choose what infrastructure is worthy of investment, and what is worthy of abandonment.

- **Shifting our thinking.** In direct contrast to linear thinking, systems thinking, lateral thinking, critical thinking, and creative thinking are foundational for getting it right in virtually every aspect of navigating the Great Unraveling. Instead of siloed thinking, we need cross-disciplinary research, the application of ecological principles, and the training and input of generalists. Thinking in terms of systems is not only key for understanding where and how to intervene but also in investing limited time and resources on *multisolving* strategies.[131]

- **Redirecting and rationing resources.** Labor, energy, materials, and money are in limited supply and will be required to build systems that can persist over the long term. These resources must be strategically withdrawn from supporting

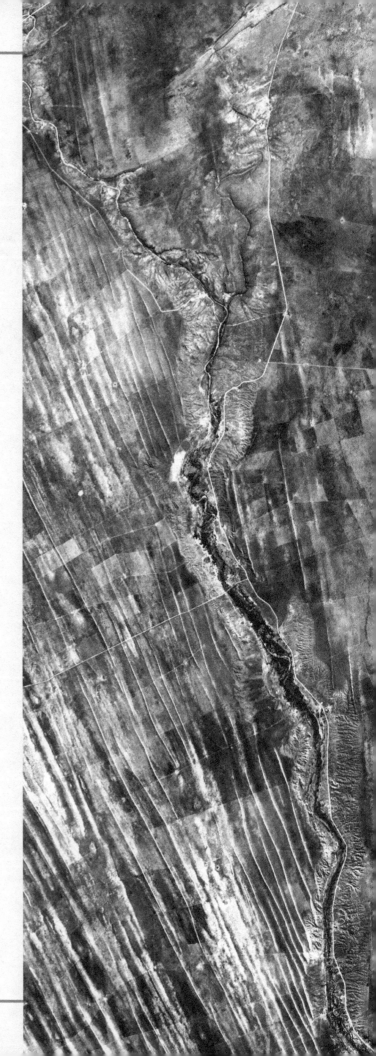

unsustainable systems and redirected toward maintaining basic services while also creating new low-energy infrastructure. And this must be done at unprecedented speed. Maintaining basic services during this wartime-scale effort will require wartime-era strategies, such as rationing of food, energy, and manufactured products. During both World Wars, rationing ensured that everyone had access to basic requirements, while Victory Gardens programs engaged a wide swath of the population in growing healthy food locally. Similar kinds of programs will be needed to minimize the human impacts of the Great Unraveling.

- **Moving from exploitation to collective care.** It's commonly understood that a tremendous source of the wealth of European and European settler nations came through the exploitation of human labor and natural resources during the period of colonialism. But this exploitation didn't cease with the end of the colonial era. It's estimated that over the 25 year period between 1990-2015, the value of all embodied raw materials, energy, land, and labor appropriated from the Global South by the Global North equates to $242 trillion.[132] As we more deeply experience the unraveling of environmental and social systems, this pattern of exploitation could persist or worsen, as wealthy nations and communities seek to transition their energy systems by extracting labor, minerals, and other "resources" from Global South and Indigenous communities, while rendering disadvantaged local communities and whole regions around the globe sacrifice zones. A reorientation toward collective care is imperative, not only from the standpoint of justice, but as the only viable path towards sustainability and security. This begins with advantaged households and nations significantly downscaling their own consumption and investing in reciprocity and reparations.

- **Adopting new (or perhaps old) stories of humanity's relationship with one another and the natural world.** During the 19th century, the English-speaking world was enthralled by the rags-to-respectability novels of Horatio Alger. In the late 20th century, the "Star Trek" narrative of humanity exploring other star systems similarly engaged the collective imagination. Both were iterations of a deeper story—the story of progress—which has permeated our collective human mentality since the beginning of the Industrial Revolution. It is an attractive story because it is optimistic, and it can feel authentic because of the rapid expansion of human powers that we've witnessed in the fossil fuel era. During the Great Unraveling, very different stories will be needed. We'll need stories that situate humanity in the context of natural systems with limits, stories that recognize the intrinsic value and our dependence on the more-than-human world, and stories of healing and regeneration. Fashioning such stories is a job for creative artists as well as public figures of all kinds.

- **Learning from Indigenous communities and those on the frontlines of the**

Great Unraveling. Indigenous peoples are defined as having lived in one place for countless generations. Living in a particular place for many centuries or millennia forces people to learn the limits of local ecosystems. Indigenous peoples typically make decisions taking future generations and other species into account. The dominant industrial cultures of the world need to rapidly adopt these essential qualities of the Indigenous mindset. However, we can't learn from Indigenous wisdom while simultaneously continuing to exterminate Indigenous culture around the globe; if we wish to learn, we must earn that knowledge through persistent alliance and generosity.

- **Building capacity to respond to changing circumstances.** The public needs help understanding why change and sacrifice are necessary—especially after decades of messaging that promoted overconsumption while excusing inequality. Public education must also be directed toward helping people build skills that will be needed in turbulent times— skills such as growing food, repairing machines and infrastructure, and providing emergency care. At the same time, public messaging must promote psychological health and motivation while building social cohesion. Maintaining the credibility of such messaging will require honest and accurate reporting of facts and transparency of institutions and office holders. What's needed is essentially the opposite of propaganda, which is typically built on unequal social power and the promotion of convenient lies. Public officials must be held to high standards of truth, fairness, and empathy. Collective psychological health and social cohesion can also be served by encouraging cultural events and expressions. In our response to the Great Unraveling, there can be a major role for the arts and for cultural pioneers in creating beauty and opportunities for social connection and enjoyment.

- **Fostering a global attitude of nonviolence, fairness, and compassion.** One likely response to rapid change (brought on both by the Great Unraveling and by collective efforts to minimize it) will be conflict over access to scarce resources, and over control of public agendas and messaging. While a certain degree of conflict may be inevitable, fighting can only make problems worse. Members of the public will need ways to express their differences without violence or hostility; they will also need to see clear evidence that leaders empathize with them that resources are being distributed equitably, and that everyone is being asked to make sacrifices when needed. Nonviolence and fairness must be pursued internationally as well, for similar reasons.

Crucially, getting our response right is not about avoiding the Great Unraveling. Systemic crisis is already here and, in many ways, inevitable. But the extent of the unraveling, how much thread we have to work with to reweave social fabric, and what we make of that thread, are all still in our hands.

VI. What You Can Do

There is a great deal that leaders of industry and government can and should do in response to the Great Unraveling. But it is extremely unlikely they will do enough to prevent an array of harmful events—from climate-related disasters to economic meltdowns. That leaves a huge burden on individuals, families, and households to adapt and respond.

For individuals, it's easy to feel overwhelmed and powerless, given the scale and diversity of the challenges our species will face this century. However, each of us does have some agency to respond courageously and creatively. There are things we can do to mitigate the worst, and to seize opportunities to build a future that's sustainable, happy, and beautiful.

Of course, it is difficult to prescribe a list of tasks or activities that work for everyone, considering widely differing circumstances, capacities, motivations, and passions. The good news is that there are plenty of areas at the household and community scale where it would be valuable to expend adaptive effort. The bad news, as we have repeatedly emphasized, is there is also nothing that can be done to entirely prevent the Great Unraveling.

In general, an overarching goal of personal action should be to build *resilience*—the capacity of a system to encounter disruption and still maintain its basic structure and functions.[133]

What is it that makes a person, a community, or an ecosystem resilient? It boils down to an ability to adapt to both short-term disruption and long-term change while retaining essential aspects of identity. Resilience in an ecological system might be described in terms of general qualities like having a diversity of species and ample nutrient reserves, plus specific adaptations like the ability of certain tree seeds to survive wildfire and sprout soon afterward. Resilience in a system where humans play a role may likewise involve general qualities like diversity (say, of revenue sources) and redundancy (say, of transportation options), plus qualities related to how we make decisions, like openness to new ideas, trust, and strong social networks.

The resilience of any one system is influenced by the resilience of everything around it (for example, the resilience of a household is affected by the resilience of the larger economy—and vice versa). Naturally, we aim to build resilience in the systems we care about the most or those where we can exert the most control. But we still need to be sensitive to ways in which influence cascades down from global systems, to nations, cities, households, and individuals—and also back up those same hierarchical levels.

Once we have a rough idea of what resilience is, we can begin to apply the concept to systems by encouraging resilience-*building* characteristics and discouraging resilience-*reducing* characteristics. It will likely be useful to focus on three specific aspects of personal resilience-building capacity: informational, emotional/psychological, and practical.

A. Informational Competence

The Great Unraveling has the potential to scramble our thinking. There will be waves of problems to sort through; many voices will seek to take advantage of our informational overwhelm; and new information delivery systems will spew ersatz explanations and identify imaginary as well as real villains. As a result, blame will proliferate, and conflict will ensue. To avoid needless hostility, we'll need more understanding—the basis of compassion and peace. And gaining more understanding, in turn, will require us to do the following:

- **Learn to think more in terms of systems.** As Donella Meadows pointed out in her classic book *Thinking in Systems,* key aspects of the world can best be understood as complicated systems, with many mutually interacting parts, rather than as simple pairs of cause-effect relationships.[134] Understanding how systems work is essential to understanding why they tend to produce problems that stubbornly resist solutions. The causes of the Great Unraveling cannot be fully grasped, nor can appropriate responses be mounted, without an understanding of system dynamics.

- **Develop critical thinking skills.** In a social environment rife with conspiracy theories and propaganda, you will need to sort truth from fiction. Critical thinking is also essential to understanding current systems and their changing status, and to being able to redesign those systems to ensure a survivable response to the Great Unraveling. Base your opinions and actions on evidence. But don't just look for evidence to confirm your existing opinions; always be open to evidence that you're wrong, and to new ways of viewing and explaining the data—but test these new explanations just as rigorously. That's the essence of the scientific method.

- **Understand the root causes of the Great Unraveling.** If we don't understand why things are falling apart, our response is likely to be chaotic and even counterproductive. Without an accurate grasp of historical context, we're likely to latch onto explanations that simply blame this or that group for the situation. There already are, and no doubt will be more, widely circulated narratives about symptoms of the Great Unraveling designed to further the interests of a particular group by blaming its enemies. But the Unraveling itself cannot be blamed on a particular group; it arose from many sources and from many different people encountering a new set of circumstances. While some people will deserve to be called to account for specific actions, becoming absorbed in the blame game won't stop the Unraveling and can considerably worsen it. Keep in mind the deeper historical dynamics that led to the Great Unraveling, and try to educate others about them.

- **Help others understand.** The more people who get it, the greater the chances of maintaining social cohesion and of mounting effective responses. Some people are better equipped than others to write or speak publicly about the Great Unraveling or publish papers about it. Even if you're not a professional communicator, it will still be helpful to share your understanding of the Unraveling, its causes, and some appropriate responses within your circle of colleagues, friends, and relatives. Your overall response efforts will be more effective, and you may have better mental health prospects, if you are surrounded by people who have this understanding.

- **Explore other ways of being and knowing**—for example, Indigenous ways of thinking. Our modern mindset is unique. Some aspects of it, such as the scientific method, deserve to be preserved for future generations. But this mindset is largely conditioned by consumerism and over-specialization. Indigenous societies tended to have a more nature-based and holistic understanding of the world and humanity's place within it. Cultivate a respect and awareness of Indigenous attitudes by seeking out useful literature and personal contacts. Also, become aware of issues affecting Indigenous people in your area and help if you can.[135]

B. Emotional/Psychological Resilience

Some people bounce back from adversity relatively easily and quickly, while others dwell in feelings of depression or anger and lose the ability to enjoy life and act effectively. Psychologists have been trying to understand why for decades. Research shows that resilient people aren't carefree eternal optimists; what distinguishes them is their adoption of successful techniques to avoid or cope with crises. Among other things, these coping techniques help balance negative emotions with positive ones and maintain an underlying sense of competence.

According to studies, these four factors are critical in personal psychological resilience:

1. The ability to make realistic plans and take steps necessary to implement them,

2. A positive self-concept and confidence in one's strengths and abilities,

3. Communication and problem-solving skills, and

4. The ability to manage strong impulses and feelings.

A personality trait often identified in psychological resilience studies is *grit*, a term that refers to perseverance and the passion for long-term goals. "Gritty" people are characterized as working persistently towards challenges and maintaining effort and interest over years despite negative feedback, adversity, plateaus in progress, or failure.

Heredity and upbringing play a role in the development of grit and emotional resilience. But later in life you can still develop these qualities through effort. Detailed advice is contained in the book *Emotional Resiliency in the Era of Climate Change* by Leslie Davenport;[136] the following are some useful approaches:

- **Work on building emotional resilience capacity and widening your window/zone of tolerance.**[137] People who have experienced significant stress or trauma may find it hard to stay calm (their fight/flight response has been triggered); alternatively, they may shut down

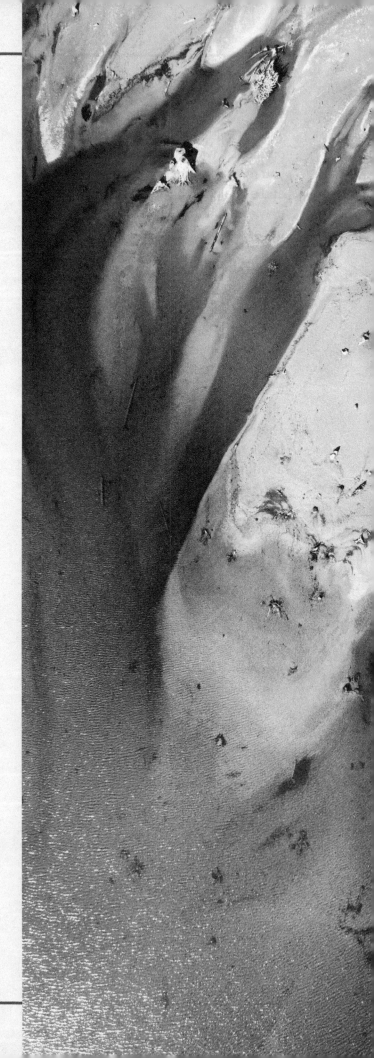

involuntarily (this is called the freeze response) when confronted with stress. Learn to become aware of the cues or signals that you are becoming either hyper- or hypo-aroused. Use these cues to alert yourself that you are coming close to the edge of your window of tolerance and that you need to take action. In general, soothing skills (such as soothing breathing and focusing on calming imagery) and grounding techniques (such as tuning into your five senses) can help bring you out of hyper-arousal back down into your optimal zone.

- **Build a centering or grounding practice.** For some, this may be meditation or prayer; for others, an artistic discipline (such as playing music, painting, or dancing), a martial art, or yoga. It should be a practice that engages both the left and right sides of the brain, and that you can pursue daily.

- **Process your feelings about the Great Unraveling.** Events in coming years—distant global events, and ones that affect you personally—will provoke feelings of grief, anger, and sadness. These are inevitable and healthy emotional responses. Find contexts in which to safely express these feelings, rather than bottling them up inside yourself.

- **Build community.** The safest space in which to process difficult feelings is a small community of people who care about one another. This could be a church, a neighborhood, or an informal network of friends. We need other people in our lives; without effort spent in building quality interpersonal connections, we tend to become increasingly isolated—and that leads to poor mental health outcomes, poor decisions, and constrained ability to respond to emergencies.

- **(Re)connect to nature and the more-than-human.** Living in urban environments walls us off from nature. As a result, we tend to lose touch with the very basis of our existence. Not only are many people ignorant of where their food comes from, but they also suffer from what psychologists call nature deficit disorder.[138] Keeping pets is one way to reconnect with the other-than-human world, but gardening and spending time in nature also fill the need to re-ground ourselves. A practical way of engaging with nature is to learn the plants and animals native to your region, and also to learn which plants are edible, which are poisonous, and which can be used for fiber and other applications.

C. Practical Personal Steps for Building Resilience

In 1989, Earth Works Group published a bestselling book titled *50 Simple Things You Can Do to Save the Earth.*[139] It advised steps like recycling, donating to environmental organizations, and eating less red meat. The book was criticized by J. Robert Hunter, author of *Simple Things Won't Save the Earth*, who argued that preserving the planet and human civilization will actually require difficult collective actions—including confronting capitalism, systemic inequality, and our societal addiction to fossil fuels.[140]

The authors of this report agree that the global polycrisis cannot be solved easily through individual efforts. Nevertheless, as we have already seen, there are things we can all do to help ourselves and our communities weather the challenges ahead, while also reducing the severity of those challenges. Here are steps to get you started:

- **Assess what vulnerabilities you face and what you can do**. What are the most likely stressors? How can you reduce your vulnerability or compensate in advance? Performing a resilience assessment requires answering a few key questions. First, *resilience of what?* What is the system you're interested in? What is its boundary? Is it your household? An ecosystem? A city? A neighborhood? An organization? If it's a human system, what are its interactions with natural systems—its resource flows, its dependencies and impacts?[141]

- **Gain skills.** Think about what skills you, your family, and colleagues may need in the face of deepening environmental and social crises. Your list might, for example, contain items like learning to grow food, or to repair machines, plumbing, and wiring. A key skill that may not come immediately to mind is conflict de-escalation and resolution. For some, self-defense may be a priority. Begin now to assemble a small relevant library, take classes, and share your new skills with others.

- **Build social cohesion and strengthen social ties.** During the COVID-19 pandemic, it became apparent that countries with higher levels of social cohesion typically saw better outcomes.[142] When people are motivated to work together, they get more done, they waste less time on bickering and complaining, and they feel better about their experience. The Great Unraveling will tear at social cohesion (as discussed above in Section III), so it is vital to nurture mutual trust and cooperation. Start now to generate connections within your neighborhood and larger community. Leadership trainings that emphasize working in groups and group psychology can be helpful. Even the simple skill of designing and running an effective meeting can make an enormous difference in enabling you to bring your community together.

- **Get to work.** Greater resilience is needed in all the essential functions and systems of every modern society—its food, money, water, waste, transport, governance, communication, security systems, and more. You will naturally gravitate toward one of these systems, given your interests, experience, and skills. Strive to increase societal resilience in your field of expertise. At the same time, seek to gain and maintain a wider and more general understanding of the status of global unraveling and the responses of various organizations and governments. If your expertise is at the national level, also pay attention to local issues affecting your community's adaptive resilience in the face of foreseeable challenges. If you are working on a resilience project, identify allies, gather resources, and make action plans with timelines. And find ways to tie your efforts with people working at national and international levels.

VII. Summary and Takeaways

The Great Unraveling is a turning point in the timeline of human existence. As such, it carries a significance on par with the emergence of language, the development of agriculture, and the Industrial Revolution. But it will likely be more perilous than those earlier watersheds. Never has humanity had so much to lose, and never has it faced so many challenges at once across so many sectors and over so short a period of time.

The Unraveling is inevitable, given the extreme and unsustainable growth trends of the past two centuries, and especially the last 70 years. Because we have adapted our collective behavior and assumptions to economic opportunities opened by vast amounts of energy unleashed via fossil fuels, we have adopted unrealistic expectations for the levels of human population and consumption that can be sustained over the long term. The dashing of those expectations against hard natural limits is one way of characterizing and understanding the Great Unraveling.

It is psychologically challenging to contemplate the unraveling of human and environmental systems. Therefore, when confronted with evidence that our current collective path is unsustainable, we tend to jump to "all-or-nothing" ways of thinking, sometimes framing our future in simplistic and unhelpful terms as "the end of the world," "apocalypse," or "collapse." While a complete and sudden end of humanity is theoretically possible via nuclear war, our more likely near-term future will consist of decades of social, economic, political, and ecological turmoil punctuated by periods of rescue and recovery. By the end of the century, both the overall human population and the overall economy will be smaller, perhaps significantly smaller, and humanity will inhabit a world of damaged but rapidly adapting ecosystems and largely depleted resources. That's not a future that many of us as individuals would willingly choose, but it is the one that we have collectively determined through decades of fossil-fueled overpopulation and overconsumption. The point to remember is that it is a future in which we will still have agency. We can optimize the Great Unraveling with cooperation and foresight, or we can ensure a worse outcome through denial and conflict.

The Unraveling has begun. Evidence discussed in Sections II and III above shows that the Unraveling is no longer merely the subject of warnings and forecasts; consequences are already unfolding and will worsen dramatically in the years and decades ahead.

Behaviors, attitudes, and strategies that seemed to make sense before the Unraveling (such as efforts to grow national and local economies) need to be replaced by different attitudes, behaviors, and strategies (such as efforts to build resilience). Building resilience at the community scale will be especially important: as global supply chains grow brittle and shatter, humanity will depend more upon local economies for survival and opportunities to thrive. Cooperative strategies to ration scarce resources and reduce inequality will also be required in order to defuse conflict and ensure optimal outcomes for as many as possible.

There is a great deal of shared hard work ahead on many levels—social, psychological, political, and practical—to minimize impacts on people and nature. To motivate this work, we must share collective goals. Our immediate goal should be to prevent harm to people and the more-than-human world while fostering resilient, diverse, ecological, nonviolent, compassionate, and more self-reliant communities. Our ultimate goal must be a way of life that offers security, fairness, and wellbeing while using energy and resources at sustainable rates and restoring natural systems rather than further degrading them.

As we work toward long-term goals, we must maximize certain short-term benefits and rewards along the way in order to maintain collective emotional health and social cohesion. By reducing inequality, by prioritizing the contributions of the creative arts, and by encouraging participatory cultural events, communities can increase their members' quality of life even when average consumption levels decline.

In responding to the Great Unraveling, we must allow and deliberately encourage some things to change, even going so far as to actively resist the forces that worsen our situation, while other things must be protected from destructive change. In the category of things that must change: people with higher incomes, both globally and within nations, will have to give up some advantages (such as easy mobility and high levels of consumption). The things we must protect include natural systems and humanity's past achievements in science, the arts, and rights. Without deliberate efforts along these lines, the Great Unraveling could leave humanity not just poorer, but culturally bereft.

If humanity descends into blame and desperate efforts to maintain a status quo that by its very nature cannot persist, the future looks dark indeed. Imagine what a young person a few decades from now, living in a depleted and ravaged world, might feel when looking at surviving images of today's "influencers" enjoying comfort, convenience, and privilege on an epic scale. What could we do now to change that scenario? Perhaps, if we work together to build a truly sustainable way of life, future generations will have some reasons to thank us.

Endnotes

1. J. R. McNeill and Peter Engelke, *The Great Acceleration: An Environmental History of the Anthropocene since 1945* (Cambridge, MA: Belknap Press, 2016).

2. Joanna Macy and Chris Johnstone, *Active Hope: How to Face the Mess We're in without Going Crazy* (Novato, CA: New World Library, 2012).

3. Union of Concerned Scientists, "1992 World Scientists' Warning to Humanity," July 16, 1992, https://www.ucsusa.org/resources/1992-world-scientists-warning-humanity; Alice Bell, "Sixty Years of Climate Change Warnings: The Signs That Were Missed (and Ignored)," *The Guardian*, July 5, 2021, https://www.theguardian.com/science/2021/jul/05/sixty-years-of-climate-change-warnings-the-signs-that-were-missed-and-ignored.

4. United Nations Environment Programme, "Rebuilding the Ozone Layer: How the World Came Together for the Ultimate Repair Job," September 15, 2021, https://www.unep.org/news-and-stories/story/rebuilding-ozone-layer-how-world-came-together-ultimate-repair-job.

5. World Economic Forum, "Global Risks Report 2020" (Geneva: World Economic Forum, 2020), https://www.weforum.org/reports/the-global-risks-report-2020/.

6. Adam Tooze, "Welcome to the World of the Polycrisis," *Financial Times*, October 28, 2022, https://www.ft.com/content/498398e7-11b1-494b-9cd3-6d669dc3de33.

7. See https://cascadeinstitute.org/research/polycrisis/ and https://omega.ngo/learn-more/the-global-polycrisis/.

8. Macy and Johnstone, *Active Hope*.

9. Michael Lawrence, Scott Janzwood, and Thomas Homer-Dixon, "What Is a Global Polycrisis?: And How Is It Different from a Systemic Risk?," Report (Cascade Institute, September 16, 2022), https://cascadeinstitute.org/technical-paper/what-is-a-global-polycrisis/.

10. National Oceanic & Atmospheric Administration, "What Is a Dead Zone?," accessed May 28, 2023, https://oceanservice.noaa.gov/facts/deadzone.html.

11. Food and Agriculture Organization of the United Nations (FAO), "The State of World Fisheries and Aquaculture 2022," Report, 2022, https://www.fao.org/publications/card/en/c/CC0461EN. Food and Agriculture Organization of the United Nations (FAO), "The State of the World's Forests 2020," Report, 2020, https://www.fao.org/publications/card/en/c/CA8642EN/.

12. Guiomar Calvo et al., "Decreasing Ore Grades in Global Metallic Mining: A Theoretical Issue or a Global Reality?," *Resources*, 2016, https://doi.org/10.3390/resources5040036.

13. World Wide Fund for Nature (WWF), "Living Planet Report 2022 – Building a Nature Positive Society," 2022, https://livingplanet.panda.org/.

14. W. Steffen et al., "The Trajectory of the Anthropocene: The Great Acceleration," *The Anthropocene Review*, 2015, https://doi.org/10.1177/2053019614564785.

15. Sinem Kilic Celik, M. Ayhan Kose, and Franziska Ohnsorge, "Can Policy Reforms Reverse the Slowing of Potential Growth?" (Brookings Institute, February 27, 2020), https://www.brookings.edu/blog/up-front/2020/02/27/can-policy-reforms-reverse-the-slowing-of-potential-growth/.

16. W. Steffen et al., "Planetary Boundaries: Guiding Human Development on a Changing Planet," *Science*, January 15, 2015, https://doi.org/10.1126/science.1259855.

17. N. J. Hagens, "Economics for the Future – Beyond the Superorganism," *Ecological Economics*, 2020, https://doi.org/10.1016/j.ecolecon.2019.106520.

18. Laura Ross, "Inside the IPhone: How Apple Sources From 43 Countries Nearly Seamlessly" (Thomas, July 21, 2020), https://www.thomasnet.com/insights/iphone-supply-chain/.

19. Peter Turchin et al., "War, Space, and the Evolution of Old World Complex Societies," *Proceedings of the National Academy of Sciences* 110, no. 41 (October 8, 2013): 16384–89, https://doi.org/10.1073/pnas.1308825110.

20. Vaclav Smil, *Energy and Civilization: A History* (Cambridge, Mass.: The MIT Press, 2018).

21. Peter Turchin, *Ultrasociety: How 10,000 Years of War Made Humans the Greatest Cooperators on Earth* (Chaplin, Ct.: Beresta Books, 2015).

22. James C. Scott, *Against the Grain: A Deep History of the Earliest States*, 1st edition (New Haven: Yale University Press, 2017).

23. Daniel Hoyer and Jenny Reddish, *Seshat History of the Axial Age* (Chaplin, CT: Beresta Books, 2019).

24. Hannah Ritchie, Max Roser, and Pablo Rosado, "Energy," *Our World in Data*, October 27, 2022, https://ourworldindata.org/energy-access.

25. Our World in Data, "Population by world region, including UN projections," accessed June 12, 2023, https://ourworldindata.org/grapher/world-population-by-region-with-projections.

26. Kerryn Higgs, "A Brief History of Consumer Culture," *The MIT Press Reader*, January 11, 2021, https://thereader.mitpress.mit.edu/a-brief-history-of-consumer-culture/.

27. Walter Scheidel, *The Great Leveler: Violence and the History of Inequality from the Stone Age to the Twenty-First Century* (Princeton, New Jersey: Princeton University Press, 2017).

28. Martin-Brehm Christensen et al., "Survival of the Richest: How We Must Tax the Super-Rich Now to Fight Inequality" (Oxfam, January 16, 2023), https://doi.org/10.21201/2023.621477.

29. Gesa Weyhenmeyer and Will Steffen, "A Warning on Climate and the Risk of Societal Collapse," *The Guardian*, December 6, 2020, sec. Environment, https://www.theguardian.com/environment/2020/dec/06/a-warning-on-climate-and-the-risk-of-societal-collapse.

30. Stephen Dovers and Colin Butler, "Population and Environment: A Global Challenge," Australian Academy of Science, accessed May 24, 2023, https://www.science.org.au/curious/earth-environment/population-environment.

31. Richard Heinberg and David Fridley, *Our Renewable Future: Laying the Path for One Hundred Percent Clean Energy* (Washington, DC: Island Press, 2016).

32. Kai Kuhnhenn et al., "A societal transformation scenario for staying below 1.5C" (Lepizig, Germany: Heinrich Böll Foundation and Konzeptwerk Neue Ökonomie 2020, December 9, 2020), https://konzeptwerk-neue-oekonomie.org/themen/degrowth/a-societal-transformation-scenario-for-staying-below-1-5c/; Joel Millward-Hopkins et al., "Providing Decent Living with Minimum Energy: A Global Scenario," Global Environmental Change 65 (November 1, 2020): 102168, https://doi.org/10.1016/j.gloenvcha.2020.102168.

33. "Ecological Footprint," Global Footprint Network, accessed May 25, 2023, https://www.footprintnetwork.org/our-

work/ecological-footprint/.

34. Johan Rockström et al., "Planetary Boundaries: Exploring the Safe Operating Space for Humanity," Ecology and Society 14, no. 2 (November 18, 2009), https://doi.org/10.5751/ES-03180-140232.

35. Steffen et al., "Planetary Boundaries: Guiding Human Development on a Changing Planet." Linn Persson et al., "Outside the Safe Operating Space of the Planetary Boundary for Novel Entities," Environmental Science & Technology 56, no. 3 (February 1, 2022): 1510–21, https://doi.org/10.1021/acs.est.1c04158.

36. William R. Catton, *Overshoot: The Ecological Basis of Revolutionary Change* (University of Illinois Press, 1982).

37. York University Ecological Footprint Initiative & Global Footprint Network. National Footprint and Biocapacity Accounts, 2022 edition. Available online at: https://data.footprintnetwork.org.

38. Corey J. A. Bradshaw et al., "Underestimating the Challenges of Avoiding a Ghastly Future," *Frontiers in Conservation Science* 1 (2021), https://www.frontiersin.org/articles/10.3389/fcosc.2020.615419.

39. European Commission Directorate-General for Climate Action, "Causes of Climate Change," accessed May 25, 2023, https://climate.ec.europa.eu/climate-change/causes-climate-change_en.

40. Valérie Masson-Delmotte et al., eds., *Climate Change 2021: The Physical Science Basis. Contribution of Working Group I to the Sixth Assessment Report of the Intergovernmental Panel on Climate Change* (Cambridge, UK and New York, NY, USA: Cambridge University Press, 2021), https://doi.org/10.1017/9781009157896.

41. World Meteorological Organization, "WMO Update: 50:50 Chance of Global Temperature Temporarily Reaching 1.5°C Threshold in next Five Years," May 9, 2022, https://public.wmo.int/en/media/press-release/wmo-update-5050-chance-of-global-temperature-temporarily-reaching-15%C2%B0c-threshold.

42. Damian Carrington, "Climate Crisis: 11,000 Scientists Warn of 'Untold Suffering,'" *The Guardian*, November 5, 2019, sec. Environment, https://www.theguardian.com/environment/2019/nov/05/climate-crisis-11000-scientists-warn-of-untold-suffering.

43. Norman G. Loeb et al., "Satellite and Ocean Data Reveal Marked Increase in Earth's Heating Rate," *Geophysical Research Letters* 48, no. 13 (2021): e2021GL093047, https://doi.org/10.1029/2021GL093047.

44. Union of Concerned Scientists, "Climate Change and Agriculture: A Perfect Storm in Farm Country," March 20, 2019, https://www.ucsusa.org/resources/climate-change-and-agriculture.

45. Kate Ramsayer, "Emissions Could Add 15 Inches to Sea Level by 2100, NASA-Led Study Finds," NASA Global Climate Change, accessed May 25, 2023, https://climate.nasa.gov/news/3021/emissions-could-add-15-inches-to-sea-level-by-2100-nasa-led-study-finds.

46. "Climate Change: More than 3bn Could Live in Extreme Heat by 2070," *BBC News*, May 5, 2020, sec. Science & Environment, https://www.bbc.com/news/science-environment-52543589.

47. Francesco Bassetti, "Environmental Migrants: Up to 1 Billion by 2050," *Foresight* (blog), May 22, 2019, https://www.climateforesight.eu/articles/environmental-migrants-up-to-1-billion-by-2050/.

48. Wood Mackenzie, "2-Degree World out of Reach Even under Accelerated Energy Transition," November 30, 2018, https://www.woodmac.com/press-releases/carbon-constrained-scenario/.

49. Andrew Glikson, "The Climate Change Runaway Chain Reaction-like Process," *Arctic News* (blog), June 20, 2021, http://arctic-news.blogspot.com/2021/06/the-climate-change-runaway-chain-reaction-like-process.html.; David Spratt and Ian

Dunlop, "Climate Dominoes: Tipping Point Risks for Critical Climate Systems" (Melbourne, Australia: Breakthrough - National Centre for Climate Restoration, May 2022), https://www.breakthroughonline.org.au/climatedominoes.

50. United Nations, "UN Report: Nature's Dangerous Decline 'Unprecedented'; Species Extinction Rates 'Accelerating,'" *United Nations Sustainable Development* (blog), May 6, 2019, https://www.un.org/sustainabledevelopment/blog/2019/05/nature-decline-unprecedented-report/.

51. Gerardo Ceballos, Paul R. Ehrlich, and Peter H. Raven, "Vertebrates on the Brink as Indicators of Biological Annihilation and the Sixth Mass Extinction," *Proceedings of the National Academy of Sciences* 117, no. 24 (June 16, 2020): 13596–602, https://doi.org/10.1073/pnas.1922686117.

52. World Wide Fund for Nature (WWF), "Living Planet Report 2022 – Building a Nature Positive Society."

53. World Economic Forum, "Global Risks Report 2020."

54. Wits University (South Africa), "Civilizations Rise and Fall on the Quality of Their Soil," *ScienceDaily*, November 4, 2013, https://www.sciencedaily.com/releases/2013/11/131104035245.htm.

55. Food and Agriculture Organization of the United Nations, "Global Soil Partnership Endorses Guidelines on Sustainable Soil Management," May 27, 2016, https://www.fao.org/global-soil-partnership/resources/highlights/detail/en/c/416516/.

56. "Torrential Rain, Flooding, and Climate Change," *SciLine- American Association for the Advancement of Science* (blog), May 27, 2020, https://www.sciline.org/climate/climate-change/torrential-rain/.

57. Pasquale Borrelli et al., "An Assessment of the Global Impact of 21st Century Land Use Change on Soil Erosion," *Nature Communications* 8, no. 1 (December 8, 2017): 2013, https://doi.org/10.1038/s41467-017-02142-7.

58. It is possible to make ammonia using electricity to obtain hydrogen from water, then combining the hydrogen with nitrogen and oxygen captured from air, but no commercially available ammonia is currently made this way. Bunro Shiozawa, "The Cost of CO2-Free Ammonia," *Ammonia Energy Association* (blog), November 12, 2020, https://www.ammoniaenergy.org/articles/the-cost-of-co2-free-ammonia/.

59. Robin McKie and Science editor, "Scientists Warn of 'Phosphogeddon' as Critical Fertiliser Shortages Loom," *The Observer*, March 12, 2023, sec. Environment, https://www.theguardian.com/environment/2023/mar/12/scientists-warn-of-phosphogeddon-fertiliser-shortages-loom.

60. Seyma Bayram, "Billions of People Lack Access to Clean Drinking Water, U.N. Report Finds," *NPR*, March 22, 2023, https://www.npr.org/2023/03/22/1165464857/billions-of-people-lack-access-to-clean-drinking-water-u-n-report-finds.

61. Alanna Shaikh, "The Bad News? The World Will Begin Running Out Of Water By 2050. The Good News? It's Not 2050 Yet," *UN Dispatch*, June 16, 2017, https://www.undispatch.com/bad-news-world-will-begin-running-water-2050-good-news-not-2050-yet/.

62. Bob Berwyn, "The Parched West Is Heading Into a Global Warming-Fueled Megadrought That Could Last for Centuries," *Inside Climate News*, April 16, 2020, https://insideclimatenews.org/news/16042020/megadrought-american-west-south-america-drought-climate-change/.

63. Mary Beth Griggs, "Two Billion People Rely On Snow For Drinking Water, And Supplies Are Melting," *Popular Science*, November 12, 2015, https://www.popsci.com/2-billion-people-rely-on-snow-for-drinking-water-and-supplies-are-melting/.

64. Jennifer Hopwood, et al., "How Neonicotinoids Can Kill Bees: The Science Behind the Role These Insecticides Play in Harming Bees" (Portland, OR: Xerces Society, 2016), https://xerces.org/publications/scientific-reports/how-neonicoti-

noids-can-kill-bees.

65. NOAA, "What Is Nutrient Pollution?," National Ocean Service, January 20, 1932, https://oceanservice.noaa.gov/facts/nutpollution.html.

66. "Fossil Fuel Air Pollution Responsible for 1 in 5 Deaths Worldwide," *C-CHANGE | Harvard T.H. Chan School of Public Health* (blog), February 9, 2021, https://www.hsph.harvard.edu/c-change/news/fossil-fuel-air-pollution-responsible-for-1-in-5-deaths-worldwide/.

67. Zeke Hausfather, "Coal in China: Estimating Deaths per GW-Year," *Berkeley Earth* (blog), November 18, 2016, https://berkeleyearth.org/deaths-per-gigawatt-year/; Health Effects Institute, "Burden of Disease Attributable to Coal-Burning and Other Air Pollution Sources in China" (Boston, Mass.: Health Effects Institute, August 18, 2016), https://www.healtheffects.org/publication/burden-disease-attributable-coal-burning-and-other-air-pollution-sources-china.

68. Ellen MacArthur Foundation, "Plastics and the Circular Economy," accessed May 26, 2023, https://ellenmacarthurfoundation.org/topics/plastics/overview.

69. Jane Muncke, "Tackling the Toxics in Plastics Packaging," *PLOS Biology* 19, no. 3 (March 30, 2021), https://doi.org/10.1371/journal.pbio.3000961.

70. Jon Hurdle, "Stealth Chemicals: A Call to Action on a Threat to Human Fertility," *Yale Environment 360*, March 18, 2021, https://e360.yale.edu/features/stealth-chemicals-a-call-to-action-on-a-threat-to-human-fertility.

71. National Institute of Environmental Health Sciences, "Perfluoroalkyl and Polyfluoroalkyl Substances (PFAS)," March 9, 2023, https://www.niehs.nih.gov/health/topics/agents/pfc/index.cfm.

72. Food and Agriculture Organization of the United Nations (FAO), "The State of World Fisheries and Aquaculture 2022"; Food and Agriculture Organization of the United Nations (FAO), "The State of the World's Forests 2020."

73. Calvo et al., "Decreasing Ore Grades in Global Metallic Mining."

74. International Energy Agency, "The Role of Critical Minerals in Clean Energy Transitions," May 14, 2021, https://www.iea.org/reports/the-role-of-critical-minerals-in-clean-energy-transitions.

75. Aurora Torres et al., "A Looming Tragedy of the Sand Commons," *Science* 357, no. 6355 (September 8, 2017): 970–71, https://doi.org/10.1126/science.aao0503.

76. Heinberg and Fridley, *Our Renewable Future*.

77. Sarah O'Connor, "Millennials Poorer than Previous Generations, Data Show," *Financial Times*, February 23, 2018, https://www.ft.com/content/81343d9e-187b-11e8-9e9c-25c814761640.

78. Ana Guinote and Theresa Vescio, eds., *The Social Psychology of Power* (New York, NY: Guilford Press, 2010), https://www.guilford.com/books/The-Social-Psychology-of-Power/Guinote-Vescio/9781606236192.

79. Karl Polanyi, *The Great Transformation: The Political and Economic Origins of Our Time*, 2nd ed. edition (Boston, MA: Beacon Press, 2001).

80. Jason Hickel et al., "Imperialist Appropriation in the World Economy: Drain from the Global South through Unequal Exchange, 1990–2015," *Global Environmental Change* 73 (March 1, 2022): 102467, https://doi.org/10.1016/j.gloenvcha.2022.102467.

81. The World Bank, "Poverty - Overview," November 30, 2022, https://www.worldbank.org/en/topic/poverty/overview.

82. The World Bank, "Poverty and Shared Prosperity 2018: Piecing Together the Poverty Puzzle" (Washington, DC, 2018), https://www.worldbank.org/en/publication/poverty-and-shared-prosperity-2018.

83. Ritchie, Roser, and Rosado, "Energy."

84. Development Initiatives, "2022 Global Nutrition Report: Strong Commitments for Greater Action" (Bristol, UK, 2022), https://globalnutritionreport.org/reports/2022-global-nutrition-report/.

85. Issac Khambule and Babalwa Siswana, "How Inequalities Undermine Social Cohesion: A Case Study of South Africa," Policy Brief (Berlin: Global Solutions Initiative, November 2, 2022), https://www.global-solutions-initiative.org/policy_brief/how-inequalities-undermine-social-cohesion-a-case-study-of-south-africa/.

86. United Nations Department of Economic and Social Affairs, "World Social Report 2020: Inequality in a Rapidly Changing World," 2020, https://www.un.org/development/desa/dspd/world-social-report/2020-2.html.

87. Inequality.org, "Global Inequality," accessed May 30, 2023, https://inequality.org/facts/global-inequality/.

88. United Nations, "Inequality – Bridging the Divide," UN75 (United Nations), accessed May 30, 2023, https://www.un.org/en/un75/inequality-bridging-divide.

89. United Nations Department of Economic and Social Affairs, "World Social Report 2020: Inequality in a Rapidly Changing World."

90. Inequality.org, "Global Inequality."

91. Anna Cooban, "Billionaires Added $5 Trillion to Their Fortunes during the Pandemic," *CNN*, January 17, 2022, https://www.cnn.com/2022/01/16/business/oxfam-pandemic-davos-billionaires/index.html.

92. International Labour Organization, "ILO Monitor: COVID-19 and the World of Work. 7th Edition" (Geneva: ILO, January 25, 2021), http://www.ilo.org/global/about-the-ilo/newsroom/news/WCMS_766949/lang--en/index.htm.

93. NIH - National Human Genome Research Institute, "Race," Genome.gov, September 14, 2022, https://www.genome.gov/genetics-glossary/Race.

94. Michael Clemens, "Why Today's Migration Crisis Is an Issue of Global Economic Inequality," Ford Foundation, July 29, 2016, https://www.fordfoundation.org/news-and-stories/stories/posts/why-today-s-migration-crisis-is-an-issue-of-global-economic-inequality/.

95. Robert Dirks et al., "Social Responses During Severe Food Shortages and Famine," *Current Anthropology* 21, no. 1 (1980): 21–44.

96. Adrian Martin, Andy Blowers, and Jan Boersema, "Is Environmental Scarcity a Cause of Civil Wars?," *Environmental Sciences* 3, no. 1 (March 1, 2006): 1–4, https://doi.org/10.1080/15693430600593981.

97. Bingham Kennedy, Jr., "Environmental Scarcity and the Outbreak of Conflict," Population Reference Bureau, accessed May 30, 2023, https://www.prb.org/resources/environmental-scarcity-and-the-outbreak-of-conflict/.

98. Steven A. LeBlanc and Katherine E. Register, *Constant Battles: Why We Fight* (New York: St. Martin's Griffin, 2004).

99. Thomas Homer-Dixon and Jessica Blitt, eds., *Ecoviolence: Links Among Environment, Population, and Security* (Lanham, MD: Rowman & Littlefield, 1998), https://rowman.com/ISBN/9780847688692/Ecoviolence-Links-Among-Environment-Population-and-Security.

100. Marianne Hanson, "Global Weapons Proliferation, Disarmament, and Arms Control," in *Global Insecurity: Futures of*

Global Chaos and Governance, ed. Anthony Burke and Rita Parker (London: Palgrave Macmillan UK, 2017), 175–93, https://doi.org/10.1057/978-1-349-95145-1_10.

101. Ted Robert Gurr, "On the Political Consequences of Scarcity and Economic Decline," *International Studies Quarterly* 29, no. 1 (1985): 51–75, https://doi.org/10.2307/2600479.

102. The Economist Intelligence Unit, "Democracy Index 2020: In Sickness and in Health?" (London, 2021), https://www.eiu.com/n/campaigns/democracy-index-2020/.

103. Rich Miller, "Rolling Zettabytes: Quantifying the Data Impact of Connected Cars," Data Center Frontier, January 21, 2020, https://www.datacenterfrontier.com/connected-cars/article/11429212/rolling-zettabytes-quantifying-the-data-impact-of-connected-cars.

104. Kate Blackwood, "Kreps: Social Media Helping to Undermine Democracy," *Cornell Chronicle*, August 20, 2020, https://news.cornell.edu/stories/2020/08/kreps-social-media-helping-undermine-democracy.

105. David Wallace, "A Large Solar Storm Could Knock out the Internet and Power Grid — an Electrical Engineer Explains How," *Astronomy Magazine*, March 21, 2022, https://www.astronomy.com/science/a-large-solar-storm-could-knock-out-the-internet-and-power-grid-an-electrical-engineer-explains-how/.

106. United Nations, "UN Analysis Shows Link between Lack of Vaccine Equity and Widening Poverty Gap," *UN News*, March 28, 2022, https://news.un.org/en/story/2022/03/1114762.

107. David Marchese, "Steven Pinker Thinks Your Sense of Imminent Doom Is Wrong," *The New York Times Magazine*, September 6, 2021, https://www.nytimes.com/interactive/2021/09/06/magazine/steven-pinker-interview.html.

108. Andrew Connelly, "A Guide to 9 Global Buzzwords for 2023, from 'polycrisis' to 'Zero-Dose Children,'" *NPR*, January 17, 2023, https://www.npr.org/sections/goatsandsoda/2023/01/17/1148994513/a-guide-to-9-global-buzzwords-for-2023-from-polycrisis-to-zero-dose-children.

109. Cascade Institute and other research institutes like the newly formed Accelerator for Systemic Risk Assessment (a special initiative of the UN Foundation) are stepping into this fold.

110. Thomas Homer-Dixon et al., "Synchronous Failure: The Emerging Causal Architecture of Global Crisis," *Ecology and Society* 20, no. 3 (July 14, 2015), https://doi.org/10.5751/ES-07681-200306.

111. IPCC, "Summary for Policymakers," in *Global Warming of 1.5°C: IPCC Special Report* (Cambridge University Press, 2022), 1–24, https://doi.org/10.1017/9781009157940.001.

112. The World Bank, "World Development Indicators | DataBank," accessed May 31, 2023, https://databank.worldbank.org/indicator/NY.GDP.MKTP.CD/1ff4a498/Popular-Indicators#.

113. Graeme S. Cumming and Garry D. Peterson, "Unifying Research on Social–Ecological Resilience and Collapse," *Trends in Ecology & Evolution* 32, no. 9 (September 1, 2017): 695–713, https://doi.org/10.1016/j.tree.2017.06.014.

114. Heinberg and Fridley, *Our Renewable Future*.

115. Thomas O. Wiedmann et al., "The Material Footprint of Nations," *Proceedings of the National Academy of Sciences* 112, no. 20 (May 19, 2015): 6271–76, https://doi.org/10.1073/pnas.1220362110.

116. Christopher Kevin Tucker, *A Planet of 3 Billion: Mapping Humanity's Long History of Ecological Destruction and Finding Our Way to a Resilient Future* (Atlas Observatory Press, 2019), 3.

117. Christopher T. M. Clack et al., "Evaluation of a Proposal for Reliable Low-Cost Grid Power with 100% Wind, Water, and

Solar," *Proceedings of the National Academy of Sciences* 114, no. 26 (June 27, 2017): 6722–27, https://doi.org/10.1073/pnas.1610381114.

118. Iñigo Capellán-Pérez, Carlos de Castro, and Luis Javier Miguel González, "Dynamic Energy Return on Energy Investment (EROI) and Material Requirements in Scenarios of Global Transition to Renewable Energies," *Energy Strategy Reviews* 26 (November 1, 2019): 100399, https://doi.org/10.1016/j.esr.2019.100399.

119. June Sekera et al., "Carbon Dioxide Removal–What's Worth Doing? A Biophysical and Public Need Perspective," *PLOS Climate* 2, no. 2 (February 14, 2023): e0000124, https://doi.org/10.1371/journal.pclm.0000124.

120. "List of Cognitive Biases and Heuristics," The Decision Lab, accessed May 31, 2023, https://thedecisionlab.com/biases.

121. Bob Joseph, "What Is the Seventh Generation Principle?," Indigenous Corporate Training, May 30, 2020, https://www.ictinc.ca/blog/seventh-generation-principle.

122. Uwe Peters, "What Is the Function of Confirmation Bias?," *Erkenntnis* 87, no. 3 (June 1, 2022): 1351–76, https://doi.org/10.1007/s10670-020-00252-1.

123. Vincent D. Costa et al., "Dopamine Modulates Novelty Seeking Behavior During Decision Making," *Behavioral Neuroscience* 128, no. 5 (October 2014): 556–66, https://doi.org/10.1037/a0037128.

124. Timothy Morton, *Hyperobjects: Philosophy and Ecology after the End of the World*, 1st edition (Minneapolis: Univ Of Minnesota Press, 2013).

125. As physicist Albert Bartlett famously quipped, "the greatest shortcoming of the human race is our inability to understand the exponential function." "Arithmetic, Population and Energy - a Talk by Al Bartlett on the Impossibility of Exponential Growth on a Finite Planet," accessed May 31, 2023, https://www.albartlett.org/presentations/arithmetic_population_energy.html.

126. Marvin Harris, *Cultural Materialism: The Struggle for a Science of Culture*, updated edition (Lanham, MD: AltaMira Press, 2001).

127. Jason Hickel and Giorgos Kallis, "Is Green Growth Possible?," *New Political Economy* 25, no. 4 (June 6, 2020): 469–86, https://doi.org/10.1080/13563467.2019.1598964.

128. Milton Friedman, *Capitalism and Freedom* (Chicago: University of Chicago Press, 1962).

129. Positive change is, of course, a subjective matter. Our view of positive change is one that leads to greater justice, equity, sustainability, and resilience.

130. Ines Perez, "Climate Change and Rising Food Prices Heightened Arab Spring," *Scientific American*, May 4, 2013, https://www.scientificamerican.com/article/climate-change-and-rising-food-prices-heightened-arab-spring/.

131. See https://www.multisolving.org/ for information and resources on multisolving.

132. Hickel et al., "Imperialist Appropriation in the World Economy."

133. "Resilience" carries different meanings in different fields. Our approach to resilience is based on how the term is understood in social-ecological systems science. For an introduction see Brian Walker and David Salt, "The Science of Resilience," *Resilience.org*, November 27, 2018, https://www.resilience.org/stories/2018-11-27/the-science-of-resilience/.

134. Donella H. Meadows, *Thinking in Systems: A Primer*, ed. Diana Wright (White River Junction, VT: Chelsea Green Publishing, 2008).

135. The Seventh Generation Fund for Indigenous Peoples is a good place to start: https://7genfund.org/.

136. Leslie Davenport, *Emotional Resiliency in the Era of Climate Change* (Philadelphia: Jessica Kingsley Pub, 2017).

137. Andreas Comninos, "Emotion Regulation 101: Your Window of Tolerance," *Mindfulness & Clinical Psychology Solutions* (blog), April 7, 2021, https://mi-psych.com.au/understanding-your-window-of-tolerance/.

138. Richard Louv, "What Is Nature-Deficit Disorder?" (Richard Louv, October 15, 2019), https://richardlouv.com/blog/what-is-nature-deficit-disorder.

139. Earth Works Group, *50 Simple Things You Can Do To Save The Earth* (Berkeley, Calif., 1989).

140. J. Robert Hunter, *Simple Things Won't Save the Earth* (Austin, Tx.: University of Texas Press, 1997).

141. There are now many resources for conducting holistic (i.e. not simply focused on disaster preparedness) resilience assessments, whether for individuals, households, or communities. Transition Network (https://transitionnetwork.org/) provides good resources and support for individuals and grassroots-led community resilience efforts. Government and institutional efforts may find the resources of Stockholm Resilience Centre (https://wayfinder.earth/) and Resilient Cities Network (https://resilientcitiesnetwork.org/) more suitable for their needs.

142. Marlon Patrick P. Lofredo, "Social Cohesion, Trust, and Government Action Against Pandemics," *Eubios Journal of Asian and International Bioethics* 30, no. 4 (2020): 182–88.

Image Credits

Cover image photo: "The Great Unraveling" by Michele Guieu; commissioned by Post Carbon Institute.

Page 1. Sediment in the Gulf of Mexico off the Louisiana coast. Image by United States Geological Survey on Unsplash.

Page 9. The Mayn River in the far northeastern corner of Siberia. Image by United States Geological Survey on Unsplash.

Page 14. The tongue of the Malaspina Glacier, the largest glacier in Alaska. Image by United States Geological Survey on Unsplash.

Page 17. Housing developments completely separated from each other in Las Vegas, Nevada. Image by NASA Earth Observatory, earthobservatory.nasa.gov

Page 20. Center-pivot irrigation in the desert in southern Egypt. Image by United States Geological Survey on Unsplash.

Page 25. Lithium mining in Salar de Atacama, Chile. Image by Coordenação-Geral de Observação da Terra/INPE, licensed under CC BY-SA 2.0; via openverse.org.

Page 31. Near the southern coast of the Netherlands. Image by United States Geological Survey on Unsplash.

Page 34. Deforestation in Bolivia. Image by NASA Earth Observatory, earthobservatory.nasa.gov

Page 38. The Mississippi River with its countless owbows and cutoffs. Image by United States Geological Survey on Unsplash.

Page 42. On the edge of the Kalahari Desert in Namibia, sand dunes encroach onto once-fertile lands. The red dot is a single center-pivot irrigation system. Image by United States Geological Survey on Unsplash.

Page 45. Phytoplankton swirl in the dark water around Gotland, Sweden. Image by United States Geological Survey on Unsplash.

Page 48. Keechelus Lake in Washington, United States. Image by iamthedave on Unsplash.

Page 51. On the edge of Great Sandy Desert in western Australia. Image by United States Geological Survey on Unsplash.

Made in the USA
Monee, IL
28 October 2023